Rockets and Raindrops

- *varying mass problems*

Craig Robinson MA(Oxon)

To my children, Gina and Conor, the most interesting and rewarding of varying mass problems.

Preface

"It will, we believe, be universally admitted that there is no easier means of becoming acquainted with any branch of Mathematics, than the study of Examples illustrative of its principles"
N M Ferrers BA (Cantab), J Stuart Jackson BA(Cantab). Preface to the Solutions of the Cambridge Senate-House Problems, for Four Years, 1848-1851

This is how *'Rockets and Raindrops'* approaches the topic of varying mass problems. There is little discussion of theory. Instead, the key areas and methods are elucidated by way of example. In this way, as seems to be often the case with mathematics, one learns through a process akin to biological osmosis, whereby there is an almost subconscious assimilation of ideas and knowledge.

Little prior understanding is assumed, and herein lies the beauty: an understanding of some basic methods of Calculus, along with Newton's Second Law of Motion, and its derivatives, is all that is required to describe the motion of a raindrop falling though a cloud, a shooting star burning up as it enters the atmosphere, or a rocket accelerating to the stars.

But the beauty lies not only in the fact that one can delineate such motion by nothing more than *'pen, paper and brain'*, but also in the actual form of the results themselves. The equations of motion that result from such *'pen, paper and brain'* exercises are beautiful in their own right, and such beauty stands the test of time.

G H Hardy FRS, 1877 – 1947, said:

"Beauty is the first test: there is no permanent place in the world for ugly mathematics."

CR, 2020

CONTENTS

1 *The Theory of Variable Mass Problems*

"I do not know what I may appear to the world, but to myself I seem to have been only like a boy playing on the sea-shore, and diverting myself in now and then finding a smoother pebble or a prettier shell than ordinary, whilst the great ocean of truth lay all undiscovered before me."
Sir Isaac Newton PRS, 1643 - 1727

1.1 <u>Variable Mass</u>

Sir Isaac Newton stated his second law of motion in this form:

"The change of motion is proportional to the motive force impressed, and is made in the direction of the right line in which the force is impressed"
Principia Mathematica, Volume I, The Motion of Bodies, Law II, Motte's Translation, Revised by Cajori

We shall interpret this as

"the rate of change of momentum of a body is proportional to the force applied"

i.e. $\dfrac{d}{dt}(mv) = F$ $\qquad\qquad\qquad\qquad$ *(1)*

This can be reformulated in the 'Momentum' form,

Impulse = Change in Momentum, or

$$Ft = mv - mu,$$

where a force, F, is applied over a time period t, resulting in a mass, m, changing velocity from u, its starting velocity, to v, its resulting velocity.

The equation (1) is equivalent to our familiar form *'F=ma"* if and only if the mass of the body concerned is constant. However, in situations where the mass of the body is not constant, we need to consider the differential $\dfrac{dm}{dt}$ and m(t), the equation giving mass after time t.

In variable mass problems, therefore, the 'Momentum' form of (1) has terms involving changing mass over small time periods.

1.2 Motion of a raindrop falling through a cloud, under the force of gravity

A raindrop, mass per unit volume m, falls through a cloud that is itself at rest. The rate at which mass is increasing is proportional to the surface area of the raindrop, which is assumed to be spherical, of radius r.
Thus

$$\frac{d}{dt}(\frac{4}{3}\pi m r^3) = 4\pi k m r^2,$$

where we have taken the constant of proportionality to be '*km*' for ease of later calculation.

The left-hand side of the above equation can be differentiated by use of the Chain Rule identity:

$$\frac{d}{dt} = \frac{d}{dr}\frac{dr}{dt},$$

and so, we have:

$$4\pi m r^2\frac{dr}{dt} = 4\pi k m r^2, \text{ or}$$

$$\frac{dr}{dt} = k$$

This gives $r = a + kt,$ *(2)*

where 'a' is the radius of the raindrop initially, at time, t=0.

If we ignore resistances, then the only force on the raindrop is its weight, and so we have, by equation (1) as above:

$$\frac{4}{3}\pi m r^3 g = \frac{d}{dt}(\frac{4}{3}\pi m r^3 v)$$

(denoting the Earth's gravitational pull by 'g' and the velocity of the raindrop by 'v')

i.e. $r^3 g = \dfrac{d}{dt}(r^3 v)$

or,

$$r^3 v = \int g r^3 dt$$

But from (2) above, we have $r = a + kt,$ and so,

$$(a + kt)^3 v = \int g(a + kt)^3 dt$$

$$= \frac{g}{4k}(a + kt)^4 + c,$$ where c is the constant of integration.

If the raindrop started from rest, i.e. v=0 when t=0, then we have

$$c = -\frac{g}{4k}a^4$$

and so,

$$(a + kt)^3 v = \frac{g}{4k}(a + kt)^4 - \frac{g}{4k}a^4$$

giving,

4

$$v = \frac{g}{4k}(a + kt) - \frac{ga^4}{4k(a+kt)^3} \qquad (3)$$

If we differentiate (3), we obtain $\dfrac{dv}{dt}$, the acceleration of the raindrop. Hence,

$$\frac{dv}{dt} = \frac{g}{4} + \frac{3ga^4}{4(a+kt)^4} \qquad (4)$$

From (4), we can see that the acceleration of the raindrop approaches $\dfrac{g}{4}$ as t becomes large, and at t=0, the acceleration of the raindrop is g, as we would expect.

1.3 Motion of a rocket

The raindrop example (1.2) is less usual. The most common, and practical, application of varying mass theory is to that of the motion of a rocket.

The acceleration of a rocket is produced by ejecting burnt fuel with a velocity (normally assumed to be constant) relative to the rocket.

In these cases, we consider the Impulse and change in Momentum over a small period of time, δt

Time = t Time = $t + \delta t$

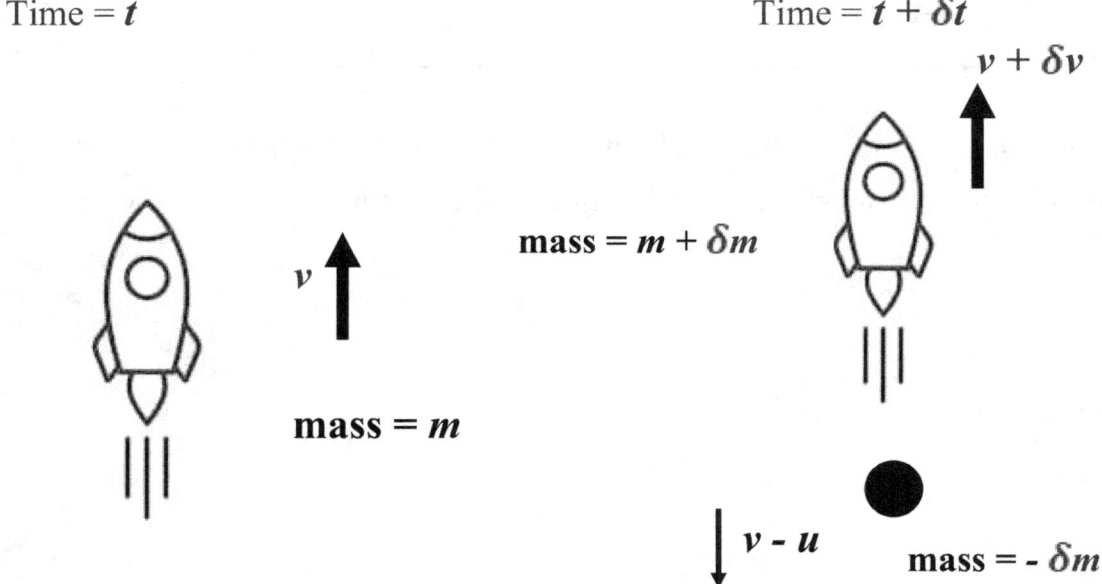

Suppose that a force, F, is acting on the rocket mass, m, velocity, v, and that the fuel is ejected with velocity, u, relative to the rocket. The mass ejected is denoted by $-\delta m$, which is a positive quantity.

Consider the change in momentum, which in general terms is equated as

$$Ft = mv - mu$$

and so, in this case we have,

$$F \, \delta t = (m + \delta m)(v + \delta v) \quad + (-\delta m)(v - u) \, - \quad mv$$

$$= m \, \delta v \, + u \, \delta m$$

neglecting the $\delta m \, \delta v$ term [we always assume that 2nd order terms can be so neglected] and setting $\delta t \to 0$, then we have

$$\boxed{F = m\frac{dv}{dt} \quad + \quad u\frac{dm}{dt}} \qquad (5)$$

If the rocket is launched vertically under the force of gravity, then F = -mg and so (5) becomes

$$\boxed{-mg = m\frac{dv}{dt} \quad + \quad u\frac{dm}{dt}} \qquad (6)$$

Let us suppose that fuel is burnt at a constant rate of say, k, then we have

$$\frac{dm}{dt} = -k$$

and solving this, we have

$$m = M_s \, + M_F \, - \, kt$$

where

M_s is the mass of the spaceship
M_F is the mass of the fuel at time t = 0

Notice that the fuel will be burnt up at time T where

$$T = \frac{M_F}{k} \qquad\qquad (6a)$$

Returning to (6), we have

$$- (M_s + M_F - kt)\, g = \frac{dv}{dt}(M_s + M_F - kt) - ku$$

giving,

$$\frac{dv}{dt} + g = \frac{ku}{M_S + M_F - kt}$$

and upon integrating both sides with respect to t, we have

$$v + gt = - u\log_e \left(\frac{M_S + M_F - kt}{M_S + M_F}\right)$$

The constant of integration has been calculated, assuming that at time t=0, we have v=0.

Notation – throughout the natural logarithm shall be denoted as $\log_e$. Readers may be used to seeing the natural logarithm also denoted as "ln".

And so, we have the equation for velocity, v, at time, t, of the rocket as

$$v = -gt + u\log_e \left(\frac{M_S + M_F}{M_S + M_F - kt} \right)$$ (7)

This shows the term "-gt" that one would expect from the usual *'suvat'* formula *"v = u + at"* (with a = -g here) applied to a field with the gravitational pull of the Earth acting as the only force, plus the extra term, which is 'positive', and which is due to the rocket propulsion.

If we look once more at the differential equation for v as above,

$$\frac{dv}{dt} + g = \frac{ku}{M_S + M_F - kt}$$

which we can write as

$$\frac{dv}{dt} = \frac{ku}{M_S + M_F - kt} - g$$

we can see for 'take off' to happen, we need $\frac{dv}{dt} > 0$ at *t =0* and that this will happen if

$$\frac{ku}{M_S + M_F} - g > 0$$

This gives us our condition for 'take off' as being

$$ku > (M_S + M_F)\, g$$ (8)

If this condition is not met, then the rocket will stay at rest until the mass has been reduced to $\dfrac{ku}{g}$ and then take off will occur because at that time the equation (8) will be satisfied. If the fuel is spent first before this mass ($\dfrac{ku}{g}$) is attained, then the rocket does not launch at all.

Now, returning to equation (7), we have

$$v = -gt + u \log_e \left(\frac{M_S + M_F}{M_S + M_F - kt} \right)$$

If we let M be $M_s + M_F$ and let s be the height attained while the rocket is burning fuel up to time T, then we have, by integrating the above expression.

(The capital T term has been used here to represent a definite time so as not to confuse with the variable, lower case t.)

$$s = -\frac{1}{2}gT^2 + u\int_o^T (\log_e(M) - \log_e(M - kt))dt$$

$$= -\frac{1}{2}gT^2 + (u\log_e(M))T - u\int_o^T \log_e(M - kt))dt$$

Put $x = M - kt$, $X = M - kT$ and so $dx = -kdt$ and, then we have

$$s = = -\frac{1}{2}gT^2 + (u\log_e(M)T + \frac{u}{k}\int_M^X \log_e x\, dx$$

and using the standard result

$$\int log_e x\,dx = xlog_e x - x,\text{ we have}$$

$$s = -\frac{1}{2}gT^2 + (ulog_e(M))T + \frac{u}{k}[xlog_e x - x]_M^X$$

$$= -\frac{1}{2}gT^2 + (ulog_e(M))T + \frac{u}{k}[\,(M\text{-}kt)log_e(M\text{-}kt) - (M\text{-}kt)]_0^T$$

$$= -\frac{1}{2}gT^2 + (ulog_e(M))T + \frac{uM}{k}log_e(M\text{-}kT) - uTlog_e(M\text{-}kT) - \frac{(M-kT)}{k}u - \frac{u}{k}Mlog_e M + \frac{u}{k}M$$

$$= -\frac{1}{2}gT^2 + uT\,[1 + log_e\frac{M}{M-kT}] + \frac{uM}{k}[log_e(M-kT) - log_e M]$$

So, the height s attained by the rocket in a general time t is

$$s = -\frac{1}{2}gt^2 + ut\,[1 + log_e(\frac{M_S + M_F}{M_S + M_F - kt})] + \frac{u(M_S + M_F)}{k}log_e(\frac{M_S + M_F - kt}{M_S + M_F})$$

Equation *(9)*

We can obtain the maximum values for v and s using equations (7) and (9) and noting that all the fuel will be burnt at time $T = \dfrac{M_F}{k}$ (equation (6a)).

Note that, for the maximum height we will need to add in a height equal to the distance travelled by the rocket whist it is still moving after the fuel has burnt out.

So,

$$V_{max} = -g\frac{M_F}{k} + uloge(\frac{M_S+M_F}{M_S+M_F-k(\frac{M_F}{k})})$$

i.e.

$$V_{max} = -g\frac{M_F}{k} + uloge(\frac{M_S+M_F}{M_S}) \tag{10}$$

and similarly, S_b, the height attained at fuel burn out is

$$S_b = -\frac{1}{2}g(\frac{M_F}{k})^2 + u\frac{M_F}{k}[1 + loge(\frac{M_S+M_F}{M_S})] + \frac{u(M_S+M_F)}{k}loge(\frac{M_S}{M_S+M_F})$$

(11)

So, the total height attained by the rocket will be S_{max} where

$$S_{max} = S_b + \frac{V_{max}^2}{2g}$$

where the second term has been calculated by using the *suvat* equations ($s = ut + (1/2)at^2$ and $v = u + at$) for a body moving under the gravitational pull of the Earth, with initial velocity V_{max} and final velocity 0 (i.e. the rocket has come to a temporary standstill at the apex of its flightpath).

Namely, from application of the first *suvat* equation

$$s = V_{max} T - \frac{1}{2}gT^2$$ where T is the time for the rocket to reach the apex of its flightpath.

But we also know that from the second *suvat* equation

$$0 = V_{max} - gT \quad \text{i.e. } T = \frac{V_{max}}{g}$$ and so from the above equation for s, we have

$$s = \frac{V_{max}^2}{g} - \frac{1}{2}g\frac{V_{max}^2}{g^2} = \frac{V_{max}^2}{2g}$$

2   *Worked Examples*

I hear and I forget. I see and I remember. I do and I understand.
Confucius, 551BC – 479BC

It is hoped that the reader will attempt these questions before recourse to the solutions as provided.

Referring back to the Preface, and its emphasis on the study of Examples, this 'study' is all the better if the mathematician attempts a solution first, before reading another mathematician's solution.

It is through the making of mistakes, and journeys down blind alleys, that one masters Mathematics. One has to **do** Mathematics in order to grasp it fully. And the Ancients knew well that this doing and understanding is re-enforced by practice

repetitio est mater studiorum
(repetition is the mother of all learning)
Old Latin proverb

Once again it can be put no better than by N M Ferrers BA (Cantab) and J Stuart Jackson BA(Cantab) in their Preface to the Solutions of the Cambridge Senate-House Problems, for Four Years, 1848-1851

"It is also indispensably necessary that the ingenuity of the Student be thoroughly exercised in attempting to discover for himself the solution of any problem which may be put before him : it is by no means our object, in publishing this book, to save him the trouble of doing so"

Model solutions are presented in a long hand form, and perhaps in doing so, some elegance is lost, but the purpose of these solutions is to impart understanding. The beauty remains.

Example 2.1

A rocket, in space, mass, including fuel, M, ejects directly behind the rocket a quantity kM of its fuel with velocity c relative to the rocket. Find the increase in velocity of the rocket, caused by this ejection of fuel.

Solution 2.1

In this case F=0 because the rocket is in the vacuum of space. The relative velocity of the fuel is -c because it is ejected behind the rocket and so the equation of motion from our equation (5) in section 1.3 as above is

$$m\frac{dv}{dt} + c\frac{dm}{dt} = 0$$

This gives

$$v = -c \int \frac{dm}{m}$$

i.e.

$$v = -c\ log_e(\frac{m}{M}) + U$$

where we have taken $v = U$ when $m=M$

i.e. U is the initial velocity.

When kM of fuel has been ejected, then, $m = (1-k)M$ and the velocity at that time, v_k, is

$$v_k = -cloge(\frac{(1-k)M}{M}) + U$$

$$= -c\log_e(1-k) + U$$

The increase in velocity is $v_k - U$, namely, $-c\log_e(1-k)$

Now

(1-k) <1 and so $\log_e(1-k)$ <0 and so $-c\log_e(1-k) > 0$

So, the increase in velocity of the rocket caused by this ejection is
$$\mid c\log_e(1-k) \mid$$

NOTE
The increase in velocity depends only on the velocity of the fuel ejected and the fraction of fuel ejected, not on the rate the fuel is consumed.

Example 2.2

A particle falls under gravity picking up rest material such that $\dfrac{dm}{dt} = mkv$, where m is its mass and v its velocity at time t. Show that

$$mg = m\frac{dv}{dt} + v\frac{dm}{dt}$$

Find the speed after falling a distance x if it starts at rest. Show that

$$x = -\frac{1}{k}\log_e(\sin\theta)$$

where $\theta = 2\,tan^{-1}(e^{-t\sqrt{gk}})$

Solution 2.2

Time = t	Force(s) acting	Time = t + δt

  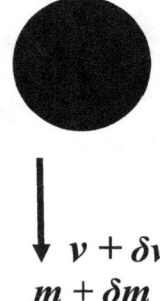

v
m

mg

$v + \delta v$
$m + \delta m$

Using 'Impulse = Change in Momentum' over the time period, **δt,** we have

$$mg\ \delta t \;=\; (m + \delta m)(v + \delta v)\ -\ mv$$

18

i.e.

$$mg\,\delta t = m\,\delta v + v\,\delta m \quad \text{ignoring the 2nd order term } \delta m\,\delta v$$

Set $\delta t \rightarrow 0$ and we have

$$mg = m\frac{dv}{dt} + v\frac{dm}{dt} \qquad \text{as required} \qquad (1)$$

Now, we are given that

$$\frac{dm}{dt} = mkv$$

and so, by (1), we have

$$mg - mkv^2 = m\frac{dv}{dt}$$

i.e

$$\frac{dv}{dt} + kv^2 = g \qquad\qquad (2)$$

Now, also we know that

$$\frac{dv}{dt} = \frac{dx}{dt}\frac{dv}{dx} \quad \text{and also, that } \frac{dx}{dt} = v$$

So,

$$\frac{dv}{dt} = v\frac{dv}{dx}$$

And returning to (2) above we therefore have

$$v\frac{dv}{dx} + kv^2 = g$$

Or

$$2v\frac{dv}{dx} + 2kv^2 = 2g \qquad\qquad (2a)$$

Either formally using the Integrating Factor technique for the solution of First Order Differential Equations or by noticing that

$$\frac{d}{dx}(e^{2kx}v^2) = e^{2kx}(2v\frac{dv}{dx} + 2kv^2)$$

we have

$$\frac{d}{dx}(e^{2kx}v^2) = 2ge^{2kx} \qquad\qquad by\ (2a)$$

and so

$$e^{2kx}v^2 = \int 2ge^{2kx}\ dx$$

$$= \frac{g}{k}e^{2kx} + c$$

But, $x=0$, when $v=0$ and so $c = \dfrac{-g}{k}$

Giving

$$e^{2kx}v^2 \;\; = \;\; \frac{g}{k}(e^{2kx} - 1)$$

i.e.

$$v^2 \;\; = \;\; \frac{g}{k}(1 - e^{-2kx})$$

Giving the velocity at distance x as [question requirement]

$$\boxed{v \;\; = \;\; \sqrt{\frac{g}{k}(1\text{-}e^{-2kx})}} \hspace{3cm} (3)$$

By (3), we have

$$\frac{dx}{dt} \;\; = \;\; \sqrt{\frac{g}{k}(1\text{-}e^{-2kx})}$$

and so, separating the variables gives

$$\int_0^x \frac{dx}{\sqrt{(1-e^{-2kx}}} \;\; = \;\; \sqrt{\frac{g}{k}} \int_0^t dt \hspace{3cm} (3a)$$

The left-hand side can be written as $\displaystyle\int_0^x \frac{e^{kx}dx}{\sqrt{(e^{kx})^2-1}}$

And now we use the substitution $u = e^{kx}$ giving $du = ke^{kx}\,dx$ and so the integral above becomes

$$\frac{1}{k}\int_0^{e^{kx}} \frac{du}{\sqrt{u^2-1}}$$

And looking back to (3a), we therefore have

$$\sqrt{\frac{g}{k}}\ \int_0^t dt \quad = \quad \frac{1}{k}\int_0^{e^{kx}} \frac{du}{\sqrt{u^2-1}}$$

Namely,

$$\sqrt{gk}\ t \quad = \quad \int_0^{e^{kx}} \frac{du}{\sqrt{u^2-1}}$$

But we have the result from Integral Calculus that

$$\int \frac{dy}{\sqrt{y^2-1}} = \log_e(y + \sqrt{y^2-1})$$

And so, by taking exponents on both sides, we have, inserting $u = e^{kx}$

$$e^{\sqrt{gk}\,t} \quad = e^{kx} \quad + \sqrt{e^{2kx}-1}$$

Now for ease of argument put $z = e^{kx}$

and the line above becomes

$$e^{\sqrt{gk}\,t} - z = \sqrt{z^2 - 1}$$

$$(e^{\sqrt{gk}\,t} - z)^2 = z^2 - 1$$

or

$$z^2 - 2ze^{\sqrt{gk}\,t} + e^{2\sqrt{gk}\,t} = z^2 - 1$$

giving

$$(1 + e^{2\sqrt{gk}\,t}) = 2ze^{\sqrt{gk}\,t}$$

or

$$z = \frac{1}{2}(e^{\sqrt{gk}\,t} + e^{-\sqrt{gk}\,t}) \qquad (4)$$

Now put

$$e^{-\sqrt{gk}\,t} = \tan\frac{1}{2}\theta$$

so that $\theta = 2\,tan^{-1}(e^{-\sqrt{gk}\,t})$ *(4a)*

then by above (4)

$$z = \frac{1}{2}(\cot\frac{1}{2}\theta + \tan\frac{1}{2}\theta) \qquad (5)$$

But

$$\cot\tfrac{1}{2}\,\theta + \tan\tfrac{1}{2}\,\theta \;=\; \frac{\cos\tfrac{1}{2}\theta}{\sin\tfrac{1}{2}\theta} + \frac{\sin\tfrac{1}{2}\theta}{\cos\tfrac{1}{2}\theta} \;=\; \frac{\cos^2\tfrac{1}{2}\theta + \sin^2\tfrac{1}{2}\theta}{\sin\tfrac{1}{2}\theta\cos\tfrac{1}{2}\theta} \;=\; \frac{1}{\sin\tfrac{1}{2}\theta\cos\tfrac{1}{2}\theta}$$

and so, by (5),

$$z = e^{kx} = \frac{1}{2\sin\tfrac{1}{2}\theta\cos\tfrac{1}{2}\theta} \;=\; \frac{1}{\sin\theta}$$

meaning that

$$x = \frac{1}{k}\log_e\!\left(\frac{1}{\sin\theta}\right)$$

i.e

$$\boxed{x = -\frac{1}{k}\log_e(\sin\theta) \qquad\qquad \text{as required}}$$

$$\text{where, by (4a) } \theta = 2\,\tan^{-1}(e^{-\sqrt{gk}\,t})$$

Example 2.3

A shell, of mass M, is at rest in space when it bursts into two fragments, the energy released being E, all of which is converted into Kinetic Energy. Show that, if the two fragments move in the same line, then the relative speed of the fragments after separation cannot be less than

$$2\sqrt{\frac{2E}{M}}$$

Solution 2.3

Before separation   After separation

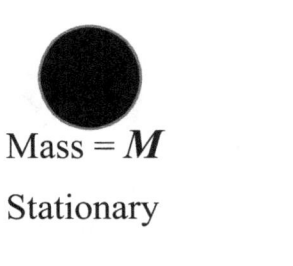   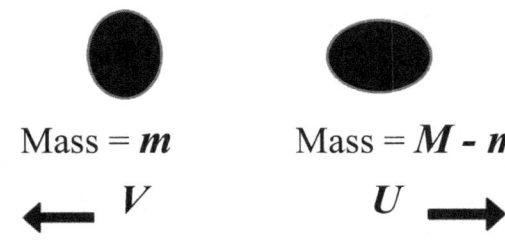

Mass $= M$ Mass $= m$ Mass $= M$ - m

Stationary $\longleftarrow V$ $U \longrightarrow$

Conservation of Momentum: $mV = (M - m)U$ (1)

Conservation of Energy: $\frac{1}{2}mV^2 + \frac{1}{2}(M-m)U^2 = E$ (2)

By (1) we have

$$m = \frac{MU}{U+V}$$

and so

$$M - m = \frac{MV}{U+V}$$

Hence by (2) we have

$$E = \frac{1}{2}\frac{MU}{U+V}V^2 + \frac{1}{2}\frac{MV}{U+V}U^2 = \frac{M}{2}\left(\frac{UV^2+VU^2}{U+V}\right) = \frac{M}{2}\frac{UV(V+U)}{U+V} = \frac{MUV}{2}$$

So,

$$UV = \frac{2E}{M} \qquad (*)$$

Now we know, in all cases, that

$$(\sqrt{U} - \sqrt{V})^2 \geq 0$$

 and so

$$U + V \geq 2\sqrt{UV}$$

But $U + V$ is precisely the relative speed of separation of the fragments, and so by (*), we have

$$U + V \geq 2\sqrt{\frac{2E}{M}} \qquad \text{as required}$$

NOTE

This example is not a standard varying mass problem in that the total mass remains the same. However, it has been included to demonstrate how the law of Conservation of Energy can sometimes be helpful in solving such problems and also because of the beauty and elegance of its solution.

It does not matter what mass the two fragments are – the relative speed of separation has a lower bound that depends only on the energy released and the total mass.

Example 2.4

A particle is projected under gravity at an angle θ. It picks up mass as $\dfrac{dM}{dt} = k$. If x is the horizontal distance and y the vertical distance and t is time, show that, if the initial mass is m,

$$x = \frac{mV\cos\theta}{k} \, log_e(1 + \frac{k}{m} t)$$

and

$$y = \frac{mV\sin\theta}{k} \, log_e(1 + \frac{k}{m} t) - \frac{1}{4} gt^2 - \frac{mg}{2k} t + \frac{m^2 g}{2k^2} log_e(1 + \frac{k}{m} t)$$

where V is its initial velocity.
Also, find the maximum height it attains.

Solution 2.4

In this diagram, h is the maximum height attained.

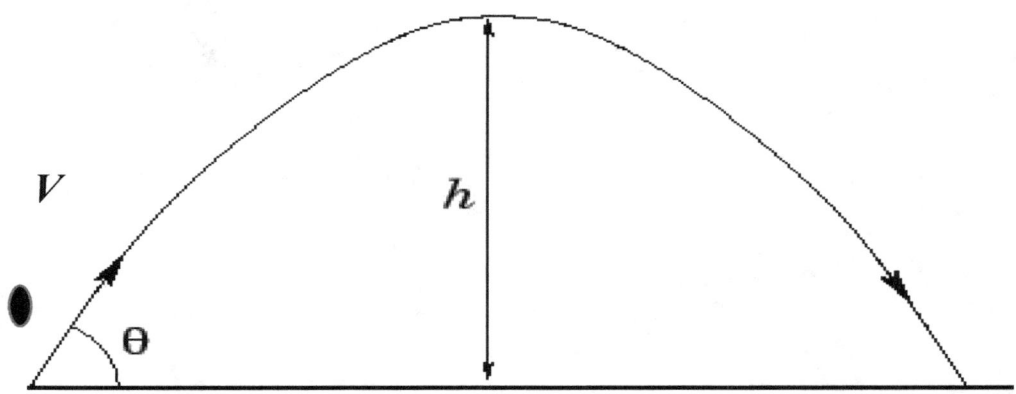

Mass = m at time t =0

We have

$$\frac{dM}{dt} = k \quad \text{and so} \quad M = m + kt \quad \text{(Mass = m at time } t = 0\text{)} \quad \text{(^)}$$

Resolve horizontally

By equation (1) in section 1.1, given that $F = 0$ horizontally, we have

$$\frac{d}{dt}(Mv) = 0 \quad \text{so that } Mv = constant$$

And so, because we know that at time $t=0$, we have horizontal momentum of $mV\cos\theta$

then
$$Mv = mV\cos\theta$$

i.e.

$$\frac{dx}{dt} = \frac{mV\cos\theta}{M} = \frac{mv\cos\theta}{m+kt} \quad \text{by (^)}$$

$$\int_0^x dx = \int_0^t \frac{mV\cos\theta}{m+kt} \, dt$$

$$= mV\cos\theta \frac{1}{k} \left[\log_e(m+kt)\right]_0^t$$

$$= \frac{mV\cos\theta}{k} \left[\log_e(m+kt) - \log_e(m)\right]$$

$$= \frac{mV\cos\theta}{k} \ log_e(\frac{(m+kt)}{m})$$

Namely,

$$x = \frac{mV\cos\theta}{k} \ log_e(1 + \frac{k}{m}t \ )$$ as required

Resolve vertically

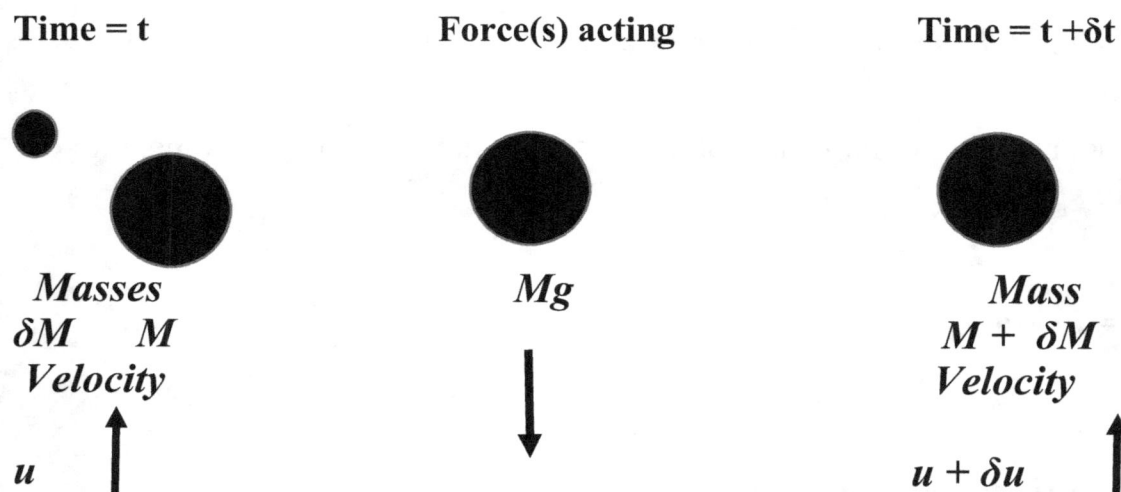

| Time = t | Force(s) acting | Time = t +δt |

Masses   **Mg**   **Mass**

δM   M   M + δM

Velocity   **Velocity**

u   u + δu

Consider Impulse =change in momentum, and we have

-Mgδt = (M + δM)(u + δu) − Mu (velocity of δM is assumed zero)

uδM + Mδu + Mgδt = 0 ignoring the **δMδu** term

i.e.

$$u\frac{dM}{dt} + M\frac{du}{dt} + Mg = 0$$

30

but

$$\frac{dM}{dt} = k \quad \text{and so} \quad M = m + kt$$

and so, we have

$$(m + kt)\frac{du}{dt} + ku = -(m + kt)g$$

or,

$$\frac{d}{dt}\big((m + kt)u\big) = -(m + kt)g$$

giving

$$(m + kt)u - mVsin\theta = \int_0^t -(m + kt)g \, dt$$

since $Vsin\theta$ is the initial upward velocity3 at time $t = 0$

Hence, we have

$$(m + kt)u - mVsin\theta = -(mt + \tfrac{1}{2}kt^2)g \qquad (1)$$

but

$$u = \frac{dy}{dt} \qquad (1a)$$

and so,

$$\frac{dy}{dt} = \frac{mV\,sin\theta}{(m+kt)} - \frac{1}{2}t\frac{(m+kt)g}{(m+kt)} - \frac{1}{2}\frac{\frac{m}{k}(m+kt)g}{(m+kt)} + \frac{1}{2}\frac{\frac{m^2}{k}}{(m+kt)}g$$

(1b)

because looking at the last three numerator terms on the right-hand side, we have

$$-\frac{1}{2}t(m+kt)g - \frac{1}{2}\frac{m}{k}(m+kt)g + \frac{1}{2}\frac{m^2}{k}g$$

$$= -\frac{1}{2}(tmg + kt^2g + \frac{m^2}{k}g + mtg - \frac{m^2}{k}g)$$

$$= -mgt - \frac{1}{2}kt^2g \quad \text{as on the right-hand side of } \textbf{\textit{(1)}}$$

And so back to (1b). we have

$$\int_0^y dy = \int_0^t \left(\frac{mV\,sin\theta}{(m+kt)} - \frac{1}{2}gt - \frac{1}{2}\frac{mg}{k} + \frac{1}{2}\frac{m^2}{k}\frac{g}{m+kt}\right)dt$$

i.e

$$y = \left[\frac{mV sin\theta}{k}\,log_e(m+kt)\right]_0^t - \frac{1}{4}gt^2 - \frac{1}{2}\frac{mgt}{k} + \left[\frac{1}{2}\frac{m^2}{k}\frac{g}{k}\frac{1}{k}\,log_e(m+kt)\right]_0^t$$

using $log_e(m+kt) - log_e(m) = log_e(\frac{m+kt}{m}) = log_e(1+\frac{k}{m}t)$ we have

$$\boxed{y = \frac{mV sin\theta}{k}\,log_e\left(1+\frac{k}{m}t\right) - \frac{1}{4}gt^2 - \frac{1}{2}\frac{mgt}{k} + \frac{m^2 g}{2k^2}\,log_e\left(1+\frac{k}{m}t\right)}$$

...................(2) as required.

The maximum height occurs when

$$u = \frac{dy}{dt} = 0 \text{ and at time, say, } T$$

Looking at equations *(1)* and *(1a)* we therefore have

$$mV\sin\theta = mTg + \frac{1}{2}kgT^2 \qquad\qquad (3)$$

This gives, *T*, the time that the maximum height is obtained as

$$T = \frac{-mg + \sqrt{m^2 g^2 + 2mkgV\sin\theta}}{kg} \qquad \text{taking the positive root.}$$

$$T = -\frac{m}{k} + \sqrt{\frac{m^2}{k^2} + \frac{2mV\sin\theta}{kg}} \qquad\qquad (3a)$$

Now we consider the non-logarithmic and logarithmic parts of equation *(2)* inserting *t = T*

Non-Logarithmic Part of (2)

$$- \left(\frac{1}{4}gT^2 + \frac{1}{2}\frac{mgT}{k} \right)$$

But by (3)

$$\frac{1}{2}kgT^2 + mgT = mV\sin\theta$$

So,

$$2k \left(\frac{1}{4} gT^2 + \frac{1}{2} \frac{mgT}{k} \right) = mV\sin\theta$$

i.e.

$$\frac{1}{4} gT^2 + \frac{1}{2} \frac{mgT}{k} = \frac{mV\sin\theta}{2k}$$

So, non-logarithmic part of equation (2) at time = T is $- \dfrac{mV\sin\theta}{2k}$ **(4)**

Logarithmic Part of (2)

At time t = T, this is

$$\left(\frac{mV\sin\theta}{k} + \frac{m^2 g}{2k^2} \right) \log_e \left(1 + \frac{k}{m} T \right)$$

$$= \frac{1}{2} \left(\frac{mV\sin\theta}{k} + \frac{m^2 g}{2k^2} \right) \log_e \left[\left(1 + \frac{k}{m} T \right)^2 \right]$$ **(4a)**

Now by **(3a)**

$$T + \frac{m}{k} = \sqrt{\frac{m^2}{k^2} + \frac{2mV\sin\theta}{kg}}$$

$$\frac{m}{k} \left(1 + \frac{k}{m} T \right) = \sqrt{\frac{m^2}{k^2} + \frac{2mV\sin\theta}{kg}}$$

$$1 + \frac{k}{m} T = \frac{k}{m} \sqrt{\frac{m^2}{k^2} + \frac{2mV\sin\theta}{kg}}$$

$$\left(1 + \frac{k}{m} T\right)^2 = \frac{k^2}{m^2}\left[\frac{m^2}{k^2} + \frac{2mV\sin\theta}{kg}\right]$$

i.e.

$$\left(1 + \frac{k}{m} T\right)^2 = 1 + \frac{2kV\sin\theta}{mg}$$

So, the logarithmic part of *(2)* is [at t = T]

$$\frac{1}{2}\left(\frac{mV\sin\theta}{k} + \frac{m^2 g}{2k^2}\right)log_e\left(1 + \frac{2kV\sin\theta}{mg}\right) \qquad\qquad \textbf{(5)}$$

Putting equations (2) (at time t=T) and (4) and (5) together we have the maximum height, h is

$$\boxed{h = \frac{1}{2}\left(\frac{mV\sin\theta}{k} + \frac{m^2 g}{2k^2}\right)log_e\left(1 + \frac{2kV\sin\theta}{mg}\right) - \frac{mV\sin\theta}{2k}}$$

NOTE:

We can perform a check on this result by setting $k \rightarrow 0$ and then comparing with the usual result for a maximum height attained for a particle, constant mass *m,* projected with velocity V at angle θ

Putting k = 0 in the above equation

We use the expansion $log_e(\,1 + x) = x - \dfrac{1}{2}x^2 + \dfrac{1}{3}x^3 \,.....$

This gives

$$h = \dfrac{1}{2}\left(\dfrac{mVsin\theta}{k} + \dfrac{m^2\,g}{2k^2}\right)\left(\dfrac{2kVsin\theta}{mg} - \dfrac{1}{2}\left(\dfrac{2kVsin\theta}{mg}\right)^2 ....\right) - \dfrac{mVsin\theta}{2k}$$

$$= \dfrac{V^2sin^2\theta}{g} + \dfrac{mVsin\theta}{2k} - \dfrac{m^2g}{4k^2}\dfrac{1}{2}\dfrac{4k^2V^2sin^2\theta}{m^2g^2} - \dfrac{mVsin\theta}{2k} +$$
order(k)

$$= \dfrac{V^2sin^2\theta}{g} - \dfrac{V^2sin^2\theta}{2g} + order(k)$$

$$= \dfrac{V^2sin^2\theta}{2g} \qquad\qquad (6)$$

setting k ➔ 0

The usual result for a maximum height attained for a particle, constant mass *m,* projected with velocity *V* at angle *θ*

Consider our original diagram and resolve vertically using the *suvat* equation "$v^2 = u^2 + 2as$"

This gives, with the final velocity being *0* at maximum height, *h*

$0 = V^2sin^2\theta - 2gh$
and so

$$h = \dfrac{V^2sin^2\theta}{2g} \qquad\qquad \text{as in equation (6) above}$$

Example 2.5

An experimental rocket engine initially of mass **m** ejects mass moving with relative velocity **u** at a rate **cm** per unit of time (**c** is a constant).

The casing of the rocket is of mass **M,** contained in the mass **m.**

Show that when the rocket is moving vertically upwards the velocity **v** at time **t** is given by

$$\frac{dv}{dt} = \frac{cu}{1-ct} - g$$

Find the condition necessary for the rocket to rise at **t = 0.**

If the rocket rises immediately show that its greatest velocity is

$$u\log_e\frac{m}{M} - \frac{g}{c}\left(1 - \frac{M}{m}\right)$$

Solution 2.5

Time = **t**

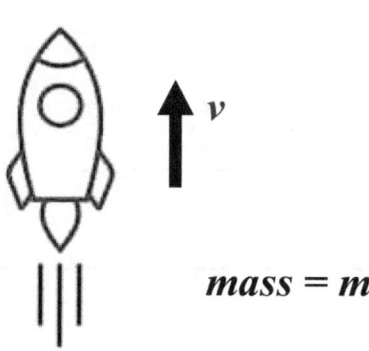

v

mass = m

Time = **t + δt**

v + δv

mass = - δm

v - u

mass = m + δm

The only force acting on the rocket is **mg** in a downward direction and so we apply

Impulse = Change in Momentum over the time period t to $t + \delta t$

$$(m + \delta m)(v + \delta v) + (-\delta m)(v - u) - mv = -mg\delta t$$

ignoring the $\delta m \, \delta v$ and setting $\delta t \rightarrow 0$

$$m\frac{dv}{dt} \; + \; u\frac{dm}{dt} + mg = 0 \qquad\qquad (1)$$

But $m(t) = m - cmt = m(1 - ct)$ $(1a)$

and so

$$\frac{dm}{dt} = -cm$$

So, equation (1) becomes

$$m(1 - ct)\frac{dv}{dt} \; - \; cmu + m(1 - ct)g \; = 0$$

Namely

$$\frac{dv}{dt} = \frac{cu}{1 - ct} - g \qquad\qquad \text{as required.} \qquad\qquad (2)$$

For take-off immediately at $t=0$ we need

$$\frac{dv}{dt} > 0 \ at \ t = 0$$

And so, we have the condition

$$\frac{cu}{1-ct} - g \ > 0 \ at \ t = 0$$

So, the condition for immediate take off is

$$cu > g \qquad\qquad (3)$$

The greatest velocity is attained when all the fuel is spent i.e. when $m(t) = M$ **at time** T

and so, by equation (1a) we have

$$M = m(1 - c \ T)$$

giving the time at which the greatest velocity is attained as

$$T = \frac{m-M}{mc} \qquad\qquad (4)$$

From (2) we have

$$\int_0^V dv \ = \int_0^T (\frac{cu}{1-ct} - g)dt$$

where V is the greatest velocity attained.

So,

$$V = \left[- u \, log_e(1- ct) - gt \right]_0^T$$

$$= -u \, log_e(1 - \frac{m-M}{m}) - g \, \frac{m-M}{mc} \quad \text{substituting in } T \text{ from equation (4)}$$

i.e.

$$V = - u \, log_e \left(\frac{M}{m}\right) - \frac{g}{c} \left(1 - \frac{M}{m} \right)$$

or

$$\boxed{V = \; u log_e\frac{m}{M} - \frac{g}{c} \left(1 - \frac{M}{m} \right)} \qquad \text{as required.}$$

NOTE

If we put $M = m$ into the equation as above

$$V = \; u log_e\frac{m}{M} - \frac{g}{c} \left(1 - \frac{M}{m} \right)$$

then we obtain

$$V = 0$$

and this is to be expected because putting $M = m$ implies that the casing of the rocket makes up the entire mass of the rocket and so there is zero mass of fuel!

Also note that in equation (2)

$$\frac{dv}{dt} = \frac{cu}{1-ct} - g$$

we have the expected "-g" term – acceleration due to gravity - and the term $\frac{cu}{1-ct}$

is the acceleration term due to the ejection of mass by the rocket engines.

Example 2.6

A rocket whose initial mass is M, half of which is accounted for by the fuel it initially contains, is fired vertically upwards. The velocity of the exhaust fumes relative to the rocket is constant and equal to u, and the mass of the fuel burnt per unit time is constant and equal to A. Assuming g is constant and neglecting air resistance show that the altitude of the rocket at which the fuel is exhausted is:

$$\frac{Mu}{2A}(1 - log_e2) \ - \ \frac{M^2g}{8A^2}$$

Solution 2.6

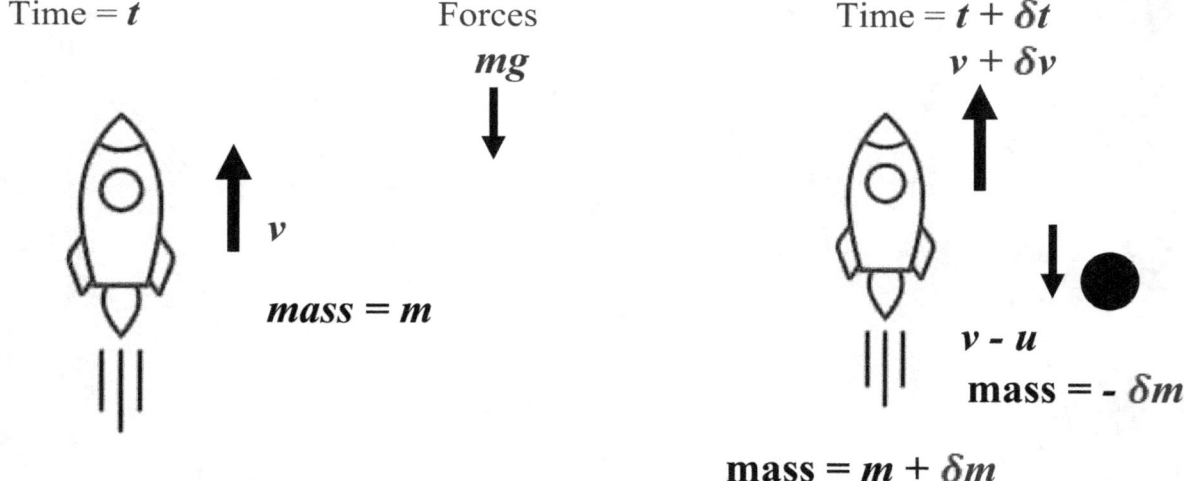

Time = t Forces   Time = $t + \delta t$

mg $v + \delta v$

v

$mass = m$

$v - u$

$mass = - \delta m$

$mass = m + \delta m$

Once again, we equate Impulse with change in momentum over the time period δt

$$(m + \delta m)(v + \delta v) + (-\delta m)(v - u) - mv = -mg\delta t$$

42

So, ignoring the $\boldsymbol{\delta m \ \delta v}$ and setting $\boldsymbol{\delta t \to 0}$

$$m\frac{dv}{dt} \ + \ u\frac{dm}{dt} + mg = 0 \qquad\qquad (1)$$

but $\ \dfrac{dm}{dt} = \text{-}A$

and so, $\ m = \dfrac{1}{2}M \ + \ \dfrac{1}{2}M \ \text{-} \ At \qquad\qquad (1a)$

and $\quad m = \dfrac{1}{2}M \quad$ when $\quad t = T = \dfrac{M}{2A} \quad (2)$

This is the time when the fuel is exhausted.

From (1) and (1a), we have:

$$(M - At)\frac{dv}{dt} \ \ \text{-} \ \ Au \ \ + (M - At)g = 0$$

and so

$$\frac{dv}{dt} \ = \ \frac{Au}{M - At} \ \text{-} \ g \quad \text{and so}$$

$$\int_0^v dv = \ \int_0^t \frac{Au}{M-At} \, dt \ \ \text{-} \ \int_0^t g \, dt$$

giving,

$$v \ = \ \text{-}u\log_e\!\left(\frac{M-At}{M}\right) \ \text{-} \ gt$$

---

i.e.

$$v = ulog_e(\frac{M}{M-At}) - gt \qquad\qquad (3)$$

Now $v = \dfrac{ds}{dt}$

At $t = T$, $s = S$ (height when fuel is exhausted) and so we have:

$$\int_0^S ds = \int_0^T (ulog_e(\frac{M}{M-At}) - gt)dt$$

and so

$$S = -\frac{1}{2}gT^2 + (ulog_eM)T - u\int_0^T (ulog_e(M - At)dt \qquad (4)$$

We now use the standard result that:

$$\int log_e x\, dx = xlog_e x - x \quad \text{and put}$$

$$z = M - At, \; dz = -Adt$$

to give

$$- u\int_0^T (ulog_e(M - At)dt = \frac{u}{A}\int_M^{M-AT} log_e z\, dz$$

$$= \frac{u}{A}\Big[zlog_e z - z\Big]_M^{M-AT}$$

$$= \frac{u}{A} \left[(M - AT) \, log_e(M - AT) \; - \; (M - AT) \; - \; M \, log_e M \; + \; M \right]$$

$$= \frac{u}{A} \left[M log_e \left(\frac{M - AT}{M} \right) \; - \; AT log_e(M - AT) \; + \; AT \right]$$

$$= \frac{uM}{A} \, log_e \left(\frac{M - AT}{M} \right) \; - \; uT \, log_e(M - AT) \; + \; uT$$

Substituting back into (4) gives –

$$S = -\frac{1}{2}gT^2 \; + \; uT \left[1 + log_e \left(\frac{M}{M - AT} \right) \right] + \frac{uM}{A} \, log_e \left(\frac{M - AT}{M} \right)$$

By (2) we have $\quad T = \dfrac{M}{2A}$

and so

$$S = \frac{1}{2}g\frac{M^2}{4A^2} \; + \; \frac{uM}{2A} \left[1 + log_e \left(\frac{M}{M - \frac{M}{2}} \right) \right] \; + \; \frac{uM}{A} \, log_e \left(\frac{M - \frac{M}{2}}{M} \right)$$

$$= \frac{-M^2 g}{8A^2} \; + \; \frac{uM}{2A} \left[1 + log_e 2 + 2 \, log_e \frac{1}{2} \right]$$

$$= \frac{-M^2 g}{8A^2} \; + \; \frac{uM}{2A} \left[1 + log_e 2 - 2 \, log_e 2 \right]$$

giving

$$S = \frac{Mu}{2A} \left[1 - \log_e 2 \right] - \frac{M^2 g}{8A^2} \qquad \text{as required}$$

NOTE

This agrees with equation *(11)* of section 1.3 as follows:

We have:

$$M_F = \frac{M}{2}$$

$$M_S = \frac{M}{2}$$

$$k = A$$

and from section 1.3 we have S as

$$S = -\frac{1}{2}g\left(\frac{M_F}{k}\right)^2 + u\frac{M_F}{k}\left[1 + \log_e\left(\frac{M_S + M_F}{M_S}\right)\right] + \frac{u(M_S + M_F)}{k}\log_e\left(\frac{M_S}{M_S + M_F}\right)$$

$$= -\frac{1}{2}g\,\frac{\frac{M^2}{4}}{A^2} + u\frac{\frac{M}{2}}{A}\left[1 + \log_e(M/(M/2))\right] + \frac{uM}{A}\log_e((M/2)/M)$$

$$= -\frac{M^2 g}{8A^2} + \frac{Mu}{2A}\left[1 + \log_e 2 + 2\log_e\frac{1}{2}\right]$$

$$= \frac{-M^2 g}{8A^2} + \frac{uM}{2A}\left[1 + \log_e 2 - 2\log_e 2\right]$$

i.e

$$S = \frac{Mu}{2A}\left[1 - \log_e 2\right] - \frac{M^2 g}{8A^2} \qquad \text{as before}$$

Example 2.7

At time t, a rocket of mass M, is in one dimensional motion with velocity V, and has a constant exhaust speed c relative to the rocket. Derive the differential equation:

$$M\frac{dV}{dt} + c\frac{dM}{dt} = F$$

where F is the total external force.

A rocket of mass M_0 of which a fraction α is fuel, carries a payload of mass m. The motor burns fuel at a constant rate $ke^{-\beta t}$ and has an exhaust speed of c. The rocket is fired vertically upwards and is subject to a constant gravitational field g.

Find the speed of the rocket as a function of time, and hence its final speed when the fuel is completely burnt.
Deduce the corresponding result for a constant fuel consumption of rate k.

Solution 2.7

Time $= t$ Forces   Time $= t + \delta t$
                        F $V + \delta V$

mass $= M + \delta M$

mass $= M$

mass $= -\delta M$

$V - c$

By considering impulse and change in momentum over the small time period δt, we have:

$$(M + \delta M)(V + \delta V) + (-\delta M)(V - c) - MV = F\delta t$$

Neglecting 2nd order terms, cancelling the MV term setting $\delta t \rightarrow 0$ we have

$$M\frac{dV}{dt} + c\frac{dM}{dt} = F \qquad \text{as required} \qquad (1)$$

In this case we have

$$\frac{dM}{dt} = -ke^{-\beta t}$$

and so, integrating and knowing that at $t=0$, $M = M_0 + m$, we have

$$\int_{M0+m}^{M} dM = \int_{0}^{t} -ke^{-\beta t}\,dt \quad = \left[\ \frac{k}{\beta}\ e^{-\beta t}\ \right]_{0}^{t}$$

Hence

$$M = M_0 + m - \frac{k}{\beta}(1 - e^{-\beta t})$$

Substituting this into equation (1) and noting that in this case $F = -Mg$, we have

$$(M_0 + m - \frac{k}{\beta}(1 - e^{-\beta t}))\frac{dV}{dt} - cke^{-\beta t} = -(M_0 + m - \frac{k}{\beta}(1 - e^{-\beta t}))g$$

giving

$$\frac{dV}{dt} - \frac{cke^{-\beta t}}{\left(M_0 + m - \frac{k}{\beta}(1 - e^{-\beta t})\right)} = -g$$

Upon integrating we have

$$\int_0^V dV = -\int_0^t dt + \int_0^t \frac{cke^{-\beta t}}{\left(M_0 + m - \frac{k}{\beta}(1 - e^{-\beta t})\right)}dt$$

$$\boxed{V = -gt + I \qquad\qquad\qquad\qquad (2)}$$

We now need to evaluate the integral I

$$I = \int_0^t \frac{cke^{-\beta t}}{\left(M_0 + m - \frac{k}{\beta} + \frac{k}{\beta}e^{-\beta t}\right)}dt \qquad \text{multiply top and bottom by } e^{\beta t}$$

$$= \int_0^t \frac{ck}{\left(M_0 + m - \frac{k}{\beta}\right)e^{\beta t} + \frac{k}{\beta}}dt$$

Now substitute

$$u = \left(M_0 + m - \frac{k}{\beta}\right)e^{\beta t} + \frac{k}{\beta}$$

then

$$du = \beta\left(M_0 + m - \frac{k}{\beta}\right)e^{\beta t}dt$$

and so

$$I = \int_{M_0+m}^{(M_0 + m - \frac{k}{\beta})e^{\beta t} + \frac{k}{\beta}} \frac{ck}{u} \frac{du}{\beta(M_0 + m - \frac{k}{\beta})e^{\beta t}}$$

But,

$$\beta(M_0 + m - \frac{k}{\beta})e^{\beta t} = \beta(u - \frac{k}{\beta}) = \beta u - k$$

So,

$$I = \int_{M_0+m}^{(M_0 + m - \frac{k}{\beta})e^{\beta t} + \frac{k}{\beta}} \frac{ckdu}{u(\beta u - k)}$$

$$= \int_{M_0+m}^{(M_0 + m - \frac{k}{\beta})e^{\beta t} + \frac{k}{\beta}} \left[\frac{\frac{\beta}{k}}{(\beta u - k)} - \frac{\frac{1}{k}}{u} \right] ckdu$$

$$= c \int_{M_0+m}^{(M_0 + m - \frac{k}{\beta})e^{\beta t} + \frac{k}{\beta}} \left[\frac{\beta}{(\beta u - k)} - \frac{1}{u} \right] du$$

$$= c \int_{M_0+m}^{(M_0 + m - \frac{k}{\beta})e^{\beta t} + \frac{k}{\beta}} \left[\frac{1}{(u - \frac{k}{\beta})} - \frac{1}{u} \right] du$$

$$= c \left[\log_e \frac{u - \frac{k}{\beta}}{u} \right]_{M_0 + m}^{(M_0 + m - \frac{k}{\beta})e^{\beta t} + \frac{k}{\beta}}$$

And so,

$$I = c\log_e \left[\frac{(M_0 + m - \frac{k}{\beta})e^{\beta t}}{(M_0 + m - \frac{k}{\beta})e^{\beta t} + \frac{k}{\beta}} \frac{M_0 + m}{M_0 + m - \frac{k}{\beta}} \right]$$

$$= c\log_e \frac{(M_0 + m)e^{\beta t}}{(M_0 + m - \frac{k}{\beta})e^{\beta t} + \frac{k}{\beta}}$$

Substituting our expression for I into equation (2), we have

$$V = -gt + c\log_e \frac{(M_0 + m)e^{\beta t}}{(M_0 + m - \frac{k}{\beta})e^{\beta t} + \frac{k}{\beta}} \quad \text{our required expression for } V$$

(3)

This is maximum when all the fuel is burnt. This happens when

$$M = M_0 + m - \alpha M_0$$

as we are told that a fraction α of the mass of the rocket is fuel.

But,

$$M = M_0 + m - \frac{k}{\beta}(1 - e^{-\beta T})$$

where T is the time at which the fuel is burnt up.

So,

$$\alpha M_0 = \frac{k}{\beta}(1 - e^{-\beta T})$$
$$\frac{\beta \alpha M_0}{k} = 1 - e^{-\beta T}$$

$$e^{-\beta T} = 1 - \frac{\beta \alpha M_0}{k}$$

$$= \frac{k - \beta \alpha M_0}{k}$$

and so

$$e^{\beta T} = \frac{k}{k - \beta \alpha M_0} \qquad\qquad (3a)$$

Now, to find V_{max} we simply substitute (3a) into equation (3) to give

$$V_{max} = -g\frac{1}{\beta} \log_e \left(\frac{k}{k - \beta \alpha M_0} \right) + c\log_e \frac{(M_0 + m)\left(\frac{k}{k - \beta \alpha M_0} \right)}{(M_0 + m - \frac{k}{\beta})\left(\frac{k}{k - \beta \alpha M_0} \right) + \frac{k}{\beta}} \qquad (4)$$

noting that $T = \frac{1}{\beta} \log_e \left(\frac{k}{k - \beta \alpha M_0} \right)$ from equation (3a)

For the case where $\beta = 0$ we have

$$\frac{dM}{dt} = -k \qquad \text{i.e} \qquad M = M_0 + m - kt \qquad (4a)$$

Now consider equation (3) which we can write as

$$V = -gt + c\log_e \frac{(M_0 + m)}{(M_0 + m) - \frac{k}{\beta}(1 - e^{-\beta t})} \qquad\qquad (5)$$

The maximum happens at time T_1 when

$$M_0 + m - \alpha M_0 = M_0 + m - kT_1 \qquad \text{by equation (4a)}$$

So, $T_1 = \dfrac{\alpha M0}{k}$ **(6)**

And so, substituting $\beta = 0$ and **(6)** into **(5)** will yield the maximum velocity for this case.

Before doing so, note that

$$\frac{k}{\beta}\left(1 - e^{-\beta t}\right) = \frac{k}{\beta}\left(1 - \left(1 - \frac{\beta t}{1!} + \frac{\beta^2 t^2}{2!} - \quad\right)\right)$$

$$= kt + \text{terms involving } \beta$$

so, as $\beta \to 0$ we have

$$\frac{k}{\beta}\left(1 - e^{-\beta t}\right) \to kt$$

and so, at the maximum velocity, noting that $T_1 = \dfrac{\alpha M0}{k}$ (by equation **(6)**),we have

$$\frac{k}{\beta}\left(1 - e^{-\beta T_1}\right) \to kT_1 = k\frac{\alpha M_0}{k} = \alpha M_0$$

Now we can substitute into equation **(5)** to give a maximum velocity in this case of

$$V_{max} = -g\frac{\alpha M0}{k} + c\log_e \frac{(M_0 + m)}{M_0 + m - \alpha M_0} \qquad \text{as was required}$$

NOTE

We can check this with equation (10) of section 1.3 which is

$$V_{max} = -g\frac{M_F}{k} + u\log_e\left(\frac{M_S+M_F}{M_S}\right) \quad (*)$$

In this case we have

$$M_S + M_F = M_0 + m$$
$$u = c$$
$$M_s = M_0 + m - \alpha M_0$$

because $M_F = \alpha M_0$

and so, *(*)* becomes

$$V_{max} = -g\frac{\alpha M0}{k} + c\log_e \frac{(M_0+m)}{M_0 + m - \alpha M_0} \quad \text{as before}$$

Example 2.8

A meteor, assumed to be spherical, of mass M, and radius a, is travelling in a straight line with constant speed V when it enters the edge of a planet's atmosphere. The meteor then experiences a resisting force of $\alpha A v^2$ where α is a constant, A the cross-sectional area and v the speed of the meteor. Friction causes the meteor to burn up so that its radius is decreasing at a rate βv where β is another constant. Mass stripped off may be assumed to be instantaneously brought to rest.

If it is assumed that gravitational forces may be neglected, find the radius of the meteor and its distance of penetration into the atmosphere as functions of time, and hence the time it takes the meteor to burn up completely and its distance of penetration at this time.

Solution 2.8

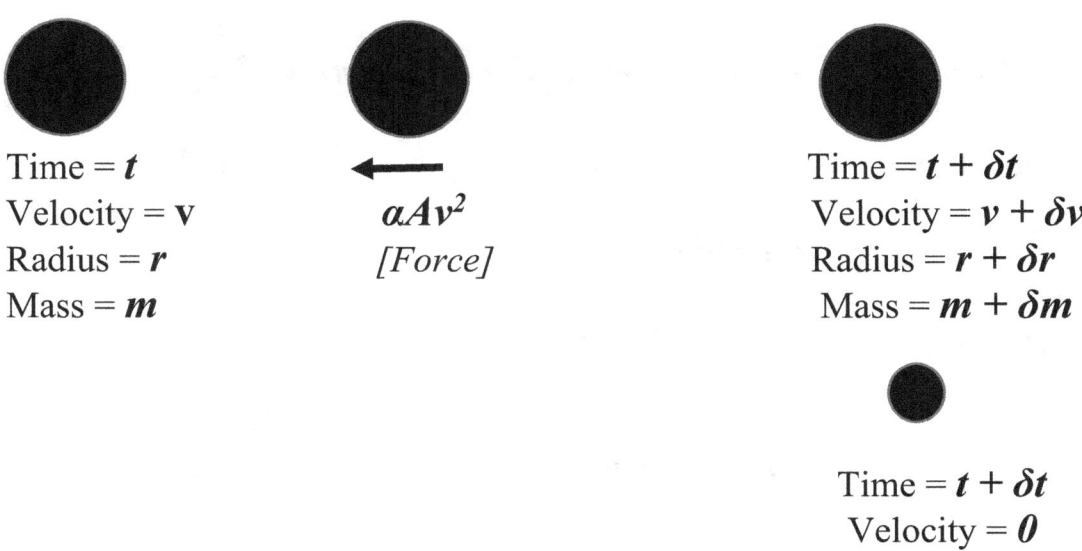

Time $= t$
Velocity $= \mathbf{v}$
Radius $= r$
Mass $= m$

$\alpha A v^2$
[Force]

Time $= t + \delta t$
Velocity $= v + \delta v$
Radius $= r + \delta r$
Mass $= m + \delta m$

Time $= t + \delta t$
Velocity $= 0$
Mass $= - \delta m$

We have that $A = \pi r^2$ and we assume a density of 1.

We are given that

$$\frac{dr}{dt} = -\beta v \qquad\qquad (1)$$

i.e.

$$\frac{dr}{dt} = -\beta\frac{dx}{dt} \qquad \text{where } x \text{ is the distance of penetration.}$$

This gives

$$r = a - \beta x \qquad\qquad (2)$$

using the initial condition that $x=0$ when $r = a$

Now, we can equate change in momentum with impulse over the time period t to $t + \delta t$

$$(m + \delta m)(v + \delta v) \quad - \delta m.0 \quad - \quad mv \quad = \quad -\alpha\,\pi r^2 v^2\,\delta t$$

Neglecting the second order term, and cancelling the mv term we have,

$$m\delta v \quad + \quad v\delta m \quad + \quad \alpha\,\pi r^2 v^2\,\delta t \quad = 0$$

Setting $\delta t \to 0$

$$m\frac{dv}{dt} \quad + \quad v\frac{dm}{dt} \quad + \quad \alpha\,\pi r^2 v^2 \ = 0 \qquad\qquad (3)$$

But

$$m = \frac{4}{3}\,\pi r^3 \quad \text{and so}$$

$$\frac{dm}{dt} = 4\,\pi r^2 \frac{dr}{dt} \qquad\qquad (4)$$

We now use the notation

$\dot{r}$ to denote $\dfrac{dr}{dt}$

The 'overdot' was Newton's notation for derivatives (which he called "fluxions" – see note below.) and you pronounce it 'r dot'.

And, likewise, we have

$\ddot{r}$ to denote $\dfrac{d^2\,r}{dx^2}$ pronounced 'r double dot'.

We also note that by (1), we have

$v = -\dot{r}/\beta$ and so, substituting (1) and (4) into (3), we have

$$-\frac{4}{3}\pi r^3\,\ddot{r}\,\frac{1}{\beta} \;-\; \frac{1}{\beta}(\dot{r})^2 4\,\pi r^2 \;+\; \frac{1}{\beta^2}\,\alpha\,\pi r^2(\dot{r})^2 \;=\; 0$$

and so, dividing by $\pi r^2/\beta$

$$\frac{4}{3}r\,\ddot{r} \;=\; (\alpha/\beta \;-\; 4)\,(\dot{r})^2$$

Now we divide by $r\dot{r}$ spotting that if we do so, then we will obtain forms on both sides of the equation that are easily integrable

---

57

$$\frac{4}{3}\frac{\ddot{r}}{\dot{r}} = (\alpha/\beta - 4)\frac{\dot{r}}{r}$$

We can make use of the standard result that

$$\int \frac{f'(x)}{f(x)}dx = \log_e(f(x)) + c$$

so that,

$$\frac{4}{3}\left[\log_e(\dot{r}) \right]_{-\beta V}^{\dot{r}} = \frac{\alpha-4\beta}{\beta}\left[\log_e(r) \right]_{a}^{r}$$

Note that when $r = a$, ie when $x=0$ (by (2)), we have $v = V$ and so $\dot{r} = -\beta V$ as given in the question rubric *"Friction causes the meteor to burn up so that its radius is decreasing at a rate βv where β is another constant."*

Evaluating the definite integrals, we have

$$\frac{4}{3} \log_e\left(\frac{\dot{r}}{-\beta V}\right) = \frac{\alpha-4\beta}{\beta}\log_e\left(\frac{r}{a}\right)$$

$$\log_e\left(\frac{\dot{r}}{-\beta V}\right) = \frac{3(\alpha-4\beta)}{4\beta} \log_e\left(\frac{r}{a}\right)$$

and so,

$$\dot{r} = -\beta V\left(\frac{r}{a}\right)^{\phi} \qquad (*)$$

where $\phi = \dfrac{3(\alpha - 4\beta)}{4\beta}$

Now we can integrate (*) to obtain an expression for r in terms of t

$$\int_a^r r^{-\phi}\, dr \;=\; \int_0^t \frac{-\beta\, V}{a^\phi}\, dt$$

Noting that $-\phi + 1 = (16\beta - 3\,\alpha)/4\beta$, we have

$$r^{(16\beta - 3\,\alpha)/4\beta} \;-\; a^{(16\beta - 3\,\alpha)/4\beta} \;=\; -\left(\frac{(16\beta - 3\,\alpha)}{4\beta}\right)\frac{\beta V}{a^\phi}\, t$$

Namely,

$$\boxed{\; r \;=\; \left[\; a^{(16\beta - 3\,\alpha)/4\beta} \;-\; \left(\frac{(16\beta - 3\,\alpha)}{4}\right)\frac{V}{a^\phi}\, t \;\right]^{4\beta/(16\beta - 3\alpha)} \quad (\$)\;}$$

This is our equation for r in terms of t.

The equation for x, the distance of penetration, in terms of t can be found by substituting the above expression for r into equation (2) above, i.e.

$$\boxed{\; x = \frac{1}{\beta}(a - r) \;}$$

Now the meteor has burned up completely when $r = 0$, and this happens when

$$x = a/\beta$$

and, in terms of when this happens, the term in the bracket on the right-hand side of *($)* is zero, at time, say, *T*, namely

$$a^{(16\beta - 3\,\alpha)/4\beta} \quad - \quad \left(\frac{(16\beta - 3\,\alpha)}{4}\right)\frac{V}{a^{\phi}}\,T = 0$$

and so

$$T = \frac{4a^{\phi}a^{(16\beta - 3\,\alpha)/4\beta}}{(16\beta - 3\,\alpha)V}$$

but,

$$a^{\phi}a^{(16\beta - 3\,\alpha)/4\beta} = a^{3(\alpha - 4\beta)/4\beta}a^{(16\beta - 3\,\alpha)/4\beta} = a$$

giving

$$T = \frac{4a}{(16\beta - 3\,\alpha)V}$$

NOTES

1. As a quick check of reasonableness, one can see that the expression for *T* above is of the right dimensions, being (distance) / (distance / time), i.e. time.
2. Fluxions was a term used by Sir Isaac Newton to describe his form of a time derivative. Newton first introduced the concept in 1665 in his mathematical treatise, *Method of Fluxions and Infinite Series*.

Newton described the method of what we mathematicians now gloss over by merely stating that we are 'setting δt → 0' as follows in his second principle of fluxions:

"...supposes that quantity is infinitely divisible, or that it may (mentally at least) so far continually diminish, as at last, before it is totally extinguished, to arrive at quantities that may be called vanishing quantities, or which are infinitely little, and less than any assignable quantity"

This resulted in much philosophical debate but stood up to scrutiny remarkably well.

Example 2.9

A woman on a railway truck running on smooth level rails is throwing out mass m of sand per second in a direction always parallel to the railway tracks, and in doing so, is doing H work per second.
Prove that the velocity of the sand relative to the truck is

$$\sqrt{\frac{2gH}{m}}$$

Solution 2.9

Let M be the mass of the truck and its contents at time t, and let V, $V + \delta V$ be the velocities of the truck at times t and $t + \delta t$ and $-v$ be the velocity of the sand relative to the truck.

So, in time δt a mass m is thrown with velocity $V - v$.

Linear Momentum is conserved because no force is acting on the system and so we have

$$(M - m\delta t)\ (V + \delta V) + m\ \delta t\ (V - v) = MV$$

As always, neglecting the 2nd order term, $\delta t\ \delta V$
we have

$$M\delta V - mv\delta t = 0 \qquad\qquad (1)$$

The work done by the woman in time δt is $Hg\delta t$ and this is equal to the increase in the kinetic energy of the system, and so we have

$$Hg\delta t = \frac{1}{2}(M - m\delta t)(V + \delta V)^2 + \frac{1}{2}m\delta t\,(V - v)^2 - \frac{1}{2}MV^2$$

Ignoring 2nd order terms, we have

$$2Hg\delta t = MV^2 + 2MV\delta V - mV^2\delta t + m\delta t\,(V^2 - 2Vv + v^2) - MV^2$$

$$= 2MV\delta V - m\delta t.V^2 + m\delta t\,V^2 - 2mVv\delta t + m\delta t\,v^2$$

$$= 2MV\delta V - 2mVv\delta t + m\delta t\,v^2$$

But by (1), we have $M\delta V = mv\delta t$

and so, the first two terms above cancel out, yielding

$$2Hg\delta t = m\delta t\,v^2$$

giving

$$v = \sqrt{\frac{2Hg}{m}} \quad \text{as required.}$$

NOTE

Generally, the use of diagrams is strongly recommended but, in some examples, where the situation is simpler, one can 'jump' straight to the equations. But if in doubt, use diagrams.

Example 2.10

A steam engine is moving along a level track with speed v subject to a total driving force of F and a resistance R. You are given that

$$\frac{d}{dt}(M(t)v) \ + \mu v = F - R$$

where μ is a constant, and is the rate at which fuel is burned, and the mass of the engine and fuel at time t is $M(t)$.

The engine works at a constant rate P and the motion is resisted by a force, R, proportional to the speed, and given by $-\frac{1}{2}kv$. If the engine, of mass M when empty starts from rest with fuel of mass m and the velocity of the engine when all the fuel is used up is V, show that, if $\frac{m}{M}$ is small enough for the square to be neglected, and the fuel is used up in time T that

$$\frac{1}{2}MV^2 \approx PT$$

Solution 2.10

We have that

$$\frac{d}{dt}(M(t)v) \ + \mu v = F - R \qquad\qquad (*)$$

and also, we know that

Power = the rate at which work is done, and so if work done is W

then, $P = \dfrac{W}{t}$ *(1)*

but Work done is the measure of energy transfer when a force (F) moves an object through a distance (d)

so, $W = Fd$ **(2)**

Putting (1) and (2) together we have

$$P = \frac{Fd}{t}$$

namely
$$P = Fv \qquad\qquad \textbf{(3)}$$

We are given that
$$R = \frac{1}{2} kv \qquad\qquad \textbf{(4)}$$

From the information in the question, we also have that

$$M(t) = M + m - \mu t \qquad \textbf{(5)}$$

We now substitute (3), (4) and (5) into (*) to obtain

$$\frac{d}{dt}((M + m - \mu t)v) + \mu v = \frac{P}{v} - \frac{1}{2} kv$$

$$(M + m - \mu t)\frac{dv}{dt} - \mu v + \mu v = \frac{P}{v} - \frac{1}{2} kv$$

$$(M + m - \mu t)\frac{dv}{dt} = \frac{P}{v} - \frac{1}{2} kv$$

$$v(M + m - \mu t)\frac{dv}{dt} = P - \frac{1}{2}kv^2$$

$$v\frac{dv}{dt} + \frac{1}{2}\frac{kv^2}{(M + m - \mu t)} = \frac{P}{(M + m - \mu t)}$$

Either by formal use of an integrating factor or by noting that

$$\frac{d}{dt}(\frac{1}{2}v^2\frac{1}{(M + m - \mu t)^{k/\mu}}) = \frac{1}{(M + m - \mu t)^{k/\mu}} \; X \; Left \; Hand \; Side$$

We have

$$\frac{d}{dt}(\frac{1}{2}v^2\frac{1}{(M + m - \mu t)^{k/\mu}}) = \frac{P}{(M + m - \mu t)^{k/\mu+1}}$$

Integrating both sides gives

$$\frac{1}{2}v^2\frac{1}{(M + m - \mu t)^{k/\mu}} = \frac{1}{k}\frac{P}{(M + m - \mu t)^{k/\mu}} + c$$

Now, at $t = 0$, we have $v = 0$ and so

$$c = -\frac{1}{k}\frac{P}{(M + m)^{k/\mu}}$$

giving

$$\frac{1}{2}v^2\frac{1}{(M + m - \mu t)^{k/\mu}} = \frac{P}{k}\left[\frac{1}{(M + m - \mu t)^{k/\mu}} - \frac{1}{(M + m)^{k/\mu}}\right]$$

i.e.

$$\frac{1}{2} v^2 = \frac{P}{k} \left[1 - \frac{(M + m - \mu t)^{k/\mu}}{(M + m)^{k/\mu}} \right]$$

or, anticipating the use of infinite expansions,

$$\frac{1}{2} v^2 = \frac{P}{k} \left[1 - \frac{(M + m)^{-k/\mu}}{(M + m - \mu t)^{-k/\mu}} \right]$$

We are given that $v = V$, when $t = T$, with T being the time when the fuel is exhausted.

But by equation (5), we have trivially that $M(T) = M$ and so

$$T = \frac{m}{\mu}$$

Substituting all this into the boxed equation, we have

$$\frac{1}{2} V^2 = \frac{P}{k} \left[1 - \frac{(M + m)^{-k/\mu}}{(M)^{-k/\mu}} \right]$$

$$= \frac{P}{k} \left[1 - (1 + \frac{m}{M})^{-k/\mu} \right]$$

Using the expansion of infinite series, knowing that $\frac{m}{M}$ is small enough for the square to be neglected, we have

$$\frac{1}{2} V^2 = \frac{P}{k} \left[1 - (1 - \frac{m}{M} \frac{k}{\mu} + o(\frac{m}{M})^2) \right]$$

$$\approx \frac{P}{k} \left[1 - 1 + \frac{m}{M} \frac{k}{\mu} \right]$$

$$\approx \frac{P}{k} \frac{m}{M} \frac{k}{\mu}$$

$$\approx P \frac{m}{M} \frac{1}{\mu}$$

So,

$$\frac{1}{2} MV^2 \approx P \frac{m}{\mu}$$

but $T = \dfrac{m}{\mu}$

and so,

$$\frac{1}{2} MV^2 \approx PT \quad \text{as required}$$

<u>NOTE</u>

The left-hand side of the above approximate equality represents the kinetic energy attained by the steam engine. The right-hand side of the above approximate equality represents the work done by the steam engine (power x time – equation (1) above). So, the result of this question implies that where $\dfrac{m}{M}$ is small, the work done by the steam engine is almost all converted into kinetic energy (of the steam engine) and this is what one would expect, and hope, to be the case where the mass of fuel is small comparable with the mass of the engine.

Example 2.11

A stationary excited nucleus suddenly explodes to form two smaller particles of masses m_1 and m_2. These travel in straight lines with relative velocity V.

Show that the total kinetic energy of the system after splitting is

$$\frac{1}{2} \frac{m_1 m_2 V^2}{(m_1 + m_2)}$$

Solution 2.11

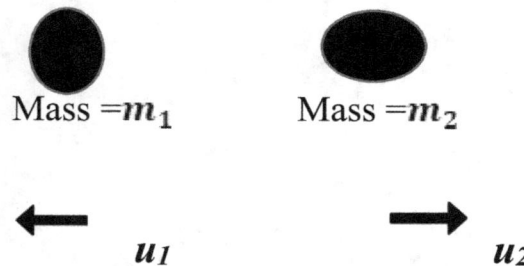

Mass $= m_1$ Mass $= m_2$

u_1 u_2

We have

Relative velocity $= V = u_1 + u_2$

Kinetic Energy $= K = \frac{1}{2} m_1 u_1^2 + \frac{1}{2} m_2 u_2^2$

By conservation of momentum : $m_1 u_1 - m_2 u_2 = 0$

So,

$2K = m_1 u_1^2 + m_2 u_2^2$

$$2 \, (m_1 + m_2) \, K = (m_1 u_1{}^2 + m_2 u_2{}^2)(m_1 + m_2)$$

$$= (m_1 u_1 - m_2 u_2)^2 + m_1 m_2 (u_1 + u_2)^2$$

but by conservation of momentum, we have, as above

$$m_1 u_1 - m_2 u_2 = 0$$

and so

$$2 \, (m_1 + m_2) \, K = m_1 m_2 (u_1 + u_2)^2$$

but

$$u_1 + u_2 = V$$

and so

$$2(m_1 + m_2) \, K = m_1 m_2 \, V^2$$

giving

$$K = \frac{1}{2} \frac{m_1 m_2 V^2}{(m_1 + m_2)} \qquad \text{as required}$$

3 *Further Problems for Solution*

"Practice makes Perfect"
English Proverb of 14th Century origins.

The reader is strongly advised to try these extra problems.

Aim to complete these problems, without reference to the notes and worked examples. Only refer backwards if you are really stuck.

It is only by doing the hard work of problem solving that Mathematics is mastered.

The aim is to achieve excellence through habit of performance and once again we look to the Ancients for their wisdom

"Excellence is an art won by training and habituation."
Aristotle, 384 BC – 322 BC

"The force of habit is great."
Cicero, 106 BC – 43 BC*

*Footnote: *Assassinated for his opposition to Mark Antony, Cicero's last words were purportedly 'There is nothing proper about what you are doing, soldier, but do try to kill me properly.'*

Problem 3.1

A raindrop of variable mass is falling through a cloud which is at rest. The mass of the raindrop increases by an amount km per unit distance of its fall. The raindrop starts from rest with mass m_0. Prove that its mass after it has fallen through a distance of x is $m_0 e^{kx}$. Obtain the equation of motion of the raindrop, and prove that its velocity v after falling through a distance x is given by

$$v^2 = \frac{g(1 - e^{-2kx})}{k}$$

Problem 3.2

A rocket has initial total mas M, including a mass m of fuel. The fuel burns at a rate of α units of mass per unit of time so as to produce a uniform thrust F. The rocket is fired vertically upwards from rest. Show that, if air resistance is neglected, the greatest velocity attained is

$$\frac{m[mg + 2(F - Mg)]}{2(M - m)\alpha}$$

Deduce that the height attained at time T $(T \leq \frac{m}{\alpha})$ is

$$\int_0^T \frac{2(F - mg) + \alpha g t}{2(M - \alpha t)} t\, dt$$

Problem 3.3

A space probe is moving free from gravitational fields, and it expels a mass m of gas per unit time from its rocket engine, with a speed a relative to the probe, where m and a are constant.
The probe experiences a resistive force equal to a constant multiple k times the square of its velocity.

Show that the equation of motion is

$$(M - mt)\frac{dv}{dt} = ma - kv^2$$

where M is the initial mass of the probe.

Hence show that if the probe is initially at rest then the velocity is related to time by the relation

$$\frac{M-mt}{M} = \left(\frac{(ma)^{\frac{1}{2}} - k^{\frac{1}{2}} v}{(ma)^{\frac{1}{2}} + k^{\frac{1}{2}} v}\right)^{(\frac{m}{4nk})^{\frac{1}{2}}}$$

Problem 3.4

A rocket is propelled vertically upwards by the backwards ejection of matter at a uniform rate and with constant velocity V relative to the rocket.
The total mass of propelling matter available is m and it is completely ejected at a time τ after launching, when the mass remaining to the rocket is km

Show that when the time is t (less than τ) the velocity of the rocket varies according to the equation

$$\frac{dv}{dt} = -g + \frac{V}{(k+1)\tau - t}$$

If $g\tau(k+1) < V$ and the initial velocity is zero, show that the rocket will rise to a height of

$$\frac{V^2}{2g}\left(\log_e \frac{k+1}{k}\right)^2 - V\tau\left[(k+1)\log_e \frac{k+1}{k} - 1\right]$$

Problem 3.5

A rocket burns fuel at a rate equal to k times its instantaneous mass, the fuel being ejected with a fixed velocity P relative to the rocket. It is initially at rest on the surface of the Earth and is fired vertically upwards.

The gravitational attraction caused by the Earth may be taken as $\frac{ga^2}{r^2}$ where a is the radius of the Earth and r is the distance of the rocket from the centre of the Earth.

Show that if $kP > g$ the relation between mass, m, and position, r, of the rocket is given by

$$log_e(m) = log_e(m_0) - \int_a^r \left(\frac{k^2 x}{2(x-a)(kPx-ag)}\right)^{1/2} dx$$

where m_o is the initial mass of the rocket.

Problem 3.6

A two-stage rocket consists of a first stage of mass M_1, when fuelled, and a second stage of mass M_2 when fuelled, and a payload of mass m. The mass of fuel in each stage is αM_1 and αM_2, respectively, where $0 < \alpha < 1.$

Each has a rocket motor which burns fuel at a mass rate k and ejects it with exhaust speed c relative to the rocket.

If the circumstances of firing are such that gravity may be neglected and $M_1 + M_2 = M$ find the optimum ratio of M_2 to M to produce the maximum final speed of payload.

Determine this maximum speed as a multiple of c if $\alpha = 0.8$ and $\frac{m}{M} = 0.01$

Problem 3.7

A rocket has an initial mass m_0 of which fraction α is fuel. The motor burns fuel at a constant rate k which is ejected with a speed c relative to the rocket. Assuming that the gravitational field is constant and that air resistance produces a retarding force of Rv, R a constant, and v the speed of the rocket, find the speed and altitude of the rocket at burnout if it is fired vertically upwards.
Calculate the altitude at burnout if
$m_o = 7000kgs$
$\alpha = 0.8$
$k = 50kgs^{-1}$
$c = 2000ms^{-1}$
$R = 4kgs^{-1}$
Is the assumption about a constant gravitational field reasonable in this case?

Hint for last part:
In a general sense, this is asking what the relationship is between gravity and the distance between masses. This is Newton's Law of Universal Gravitation given by
$F = \frac{Gm_1m_2}{r^2}$ where G is the gravitational constant (6.674×10^{-11} m^{3}kg^{-1}s^{-2}), m_1, m_2 are the masses of the 2 objects, r is the distance between them, and F, the resulting force of attraction. So, gravity is inversely proportional to the square of the distance between the interacting objects. An approximation for how gravitational pull varies with height is given by this formula
$g^* = g\left(\frac{R}{R+x}\right)^2$ where g^* is the gravitational acceleration at height x above sea level, R is the Earth's mean radius (6,378 km at the Equator to 6,357 km at the Poles) and g is the standard gravitational acceleration (9.80665 m/s^2) at the surface of the Earth. See section 4.2 for a justification of this approximation.
Section 4 makes use of Newton's Law of Universal Gravitation to calculate the escape velocity of the Earth's gravitational field.

Problem 3.8

A rocket of initial mass m_0 burns fuel at a constant rate of α units of mass per unit of time, and ejects it backwards with constant speed u relative to the rocket where $\alpha u > m_0 g$ and g is the gravitational acceleration, which may be assumed to be independent of height. The rocket is launched vertically upwards with zero initial speed, air resistance is negligible, and the total mass of fuel available to be burnt is $\frac{1}{2}m_0$. By calculating the net change in momentum over the time interval t to $t + \delta t$, or otherwise, show that the equation of motion of the rocket is

$$m\frac{dv}{dt} = \alpha u - mg$$

where m is the mass of the rocket at time t and v is its speed at that time.

Show that the speed of the rocket when all the fuel is burnt is

$$u \log_e 2 \; - \; \frac{m_0 g}{2\alpha}$$

Calculate also the *overall* height risen by the rocket.

Problem 3.9

A particle moves in a straight line under a force F, its mass increasing by picking up matter whose previous velocity was u. If the mass and velocity of the particle at time t are m and v respectively, show that

$$\frac{d}{dt}(mv) - \frac{dm}{dt}u = F$$

A particle whose mass at time t is $m(1 + \alpha t)$ is projected vertically upwards under gravity at time $t = 0$ with velocity V, the added mass being picked up from rest. Show that it rises to a height

$$\frac{g+2\alpha V}{4\alpha^2} \, log_e(1 + \frac{2\alpha V}{g}) - \frac{V}{2\alpha}$$

NOTE

Once you have attempted this question, you can compare the quoted result for the height risen to that in Example 2.4.

Problem 3.9 is a special case of Example 2.4 with $\theta = \dfrac{\pi}{2}$ and the other parameters between the two are related as such

"$m = m$"
"$k = \alpha m$"

The parameter for Example 2.4 is on the left of these two equalities and the parameter for Problem 3.9 is on the right of these two equalities.

The equation for the overall height attained in Example 2.4 is given by

$$H = \frac{1}{2}(\frac{mV sin\theta}{k} + \frac{m^2 g}{2k^2}) \, log_e \, (1 + \frac{2kV sin\theta}{mg}) - \frac{mV sin\theta}{2k}$$

and so, substituting those values as above into this formula we have, noting that $\sin(\theta = \frac{\pi}{2}) = 1$

$$H = \frac{1}{2}(\frac{mv}{\alpha m} + \frac{m^2 g}{2\alpha^2 m^2}) \ log_e \ (1 + \frac{2\alpha m V}{mg}) \quad - \quad \frac{mV}{2\alpha m}$$

i.e.

$$H = \frac{1}{2}(\frac{v}{\alpha} + \frac{g}{2\alpha^2}) \ log_e \ (1 + \frac{2\alpha V}{g}) \quad - \quad \frac{V}{2\alpha}$$

$$= \frac{g + 2\alpha V}{4\alpha^2} \ log_e(1 + \frac{2\alpha V}{g}) \quad - \quad \frac{V}{2\alpha}$$

as above.

Problem 3.10

A mass m of water issues per unit time from a pipe with uniform velocity u, and strikes a bucket, which retains all the water, there being no elasticity. Initially the bucket is at rest, and at a subsequent instant it is moving in the direction of the stream of water issuing from the pipe with velocity V. Prove that

$$\frac{dV}{dt} = \frac{m(u-V)^3}{Mu^2}$$

and also, that the loss of energy up to this instant is

$$\frac{1}{2} MuV$$

where M is the mass of the bucket, and gravity is omitted from consideration.

Problem 3.11
A body of mass $m_1 + m_2$ is split into two parts of masses m_1 and m_2 by an internal explosion which generates kinetic energy E. Show that if after separation the two parts move in a straight line then their relative speed is

$$\sqrt{\frac{2E(m_1 + m_2)}{m_1 m_2}}$$

NOTE
Compare this with Example 2.3.

We have as above

$$\text{relative speed} = \sqrt{\frac{2E(m_1 + m_2)}{m_1 m_2}} = \frac{\sqrt{2E(m_1 + m_2)}}{\sqrt{m_1 m_2}} \qquad (**)$$

But we also know that

$$(\sqrt{m_1} - \sqrt{m_2})^2 \geq 0$$

Giving

$$\frac{m_1 + m_2}{2} \geq \sqrt{m_1 m_2}$$

So, by (**)

we have that

83

relative speed $\geq \dfrac{\sqrt{2E(m_1 + m_2)}}{(\frac{m_1 + m_2}{2})}$

$$= 2\sqrt{\dfrac{2E}{m_1 + m_2}}$$

$$= 2\sqrt{\dfrac{2E}{M}} \quad \text{as in Example 2.3}$$

where $M = m_1 + m_2$

Problem 3.12

A particle of mass m and kinetic energy E is travelling uniformly in a straight line when a release of internal energy splits it into two particles, each of mass $\dfrac{m}{2}$ with total kinetic energy $E + Q$.

One of these particles is observed to be moving at right angle to the original direction of motion.

Find the speed and direction of motion of the other particle.

Show that the observed motion implies that $Q > E$

4   *Escape velocity of the Earth*

"How did I escape? With difficulty. How did I plan this moment? With pleasure."
Alexandre Dumas, 1802-1870

4.1 <u>Calculation of escape velocity of Earth</u>

As noted in Problem 3.7, Newton's Law of Universal Gravitation is given by the formula

$$F = \frac{Gm_1m_2}{r^2}$$

where G is the gravitational constant (6.674×10^{-11} m^{3}kg^{-1}s^{-2}),
m_1, m_2 are the masses of the 2 objects,
r is the distance between them, and
F, the resulting force of attraction.

If a force acting on an object is a function of position only, it is said to be a 'conservative force', and it can be represented by a Potential Energy Function, which satisfies this equation

$$U(x) := -\int_{r_0}^{r} F(x)dx$$

The $U(x)$ function is taken as the definition of Potential Energy.

There will be an arbitrary constant of integration from this integral but in practice this means that we choose things to be 'convenient' for the example that we are dealing with.

The potential energy is equal to the Work you must do to move an object from a position r_0 to a position r.

The negative sign appears in the above formula because the potential energy is equal to the work you must do against the force (exerted by a gravitational field here) to move the object from a position r_0 to a position r.

That is to say the force you must exert is equal but oppositely directed.

It is called a 'conservative force' because all paths between two arbitrary points will require the same amount of work.

Now in the case of our Newtonian Gravitational field force, we have

$$U(r) = \int \frac{Gm_1m_2}{r^2}\, dr$$

There is no minus sign in the above integral because the force of attraction due to gravitation is assumed to be 'downward' in this case.

And so

$$U(r) = -\frac{GMm}{r+R}$$

where our rocket has mass m, assumed constant for these purposes, and
where the Earth has mass M, and
where R is the radius of the Earth and
where r is the distance of the rocket above the surface of the Earth

Within the formula as above, we have effectively chosen our constant of integration. So, the formula takes the form as above because in this case we need to calculate how much work needs to be done against the force acting on the rocket, and the force of attraction between the Earth and the rocket will be calculated considering the Earth to be a point mass at distance R from the rocket, when the rocket is stationary on the surface of the Earth.

Putting $r = 0$ in the above formula yields

$$U(0) = -\frac{GMm}{R}$$

which is exactly the gravitational force acting on the rocket, stationary on the surface of the Earth.

We are now closer to calculating the escape velocity.

But, first, we must introduce the concept of Conservation of Energy, which when considering Kinetic and Potential Energy can be stated as
Kinetic Energy + Potential Energy = Constant
[see 4.3 at the end of this section]

And so, in this case, we could write

$$U(0) + K(0) = U(\infty) + K(\infty)$$

Here the **U** function represents potential energy and the **K** function represents kinetic energy and point **0** represents the surface of the Earth and ∞ represents a distance which can be taken as the rocket having escaped the Earth's gravitational field.

We want our rocket to 'just reach' 'infinity' (i.e. we want to find the minimum velocity needed to escape the Earth's gravitational field), and by 'just' we also mean that the rocket will have zero kinetic energy at that point.

So, in the above equation we set

$$U(\infty) = K(\infty) = 0$$
and therefore, we have

$$U(0) + K(0) = 0 \qquad\qquad (+)$$

Now at lift off the rocket will have kinetic energy equal to
$$\frac{1}{2}mv^2$$

and so, equation *(+)* as above becomes

$$-\frac{GMm}{R} + \frac{1}{2}mv^2 = 0$$

Giving, where we have renamed v as v_e to represent the escape velocity of the Earth

$$v_e = \sqrt{\frac{2GM}{R}}$$

In the case of the Earth we have

M = 5.974 x 10²⁴ kg
Equatorial radius, R, = 6378km = 6,378,000m

and in all cases, we have

G = 6.674×10⁻¹¹ m³kg⁻¹s⁻²

This gives

$$v_e{}^2 = \frac{2 \times 6.674 \times 10^{-11} \times 5.974 \times 10^{24}}{6,378,000} = 125,025,011$$

Giving
v_e = 11,181 m/s

So, we have the result that

$$v_e = 11.2km/s$$

Or in 'old money', $v_e \approx 25{,}000mph!$

NOTE

It is often a good idea to check that the dimensions of equations give sensible results too. In this case the dimensions on the right-hand side of this equation

$$v_e = \sqrt{\frac{2GM}{R}}$$

give

$$\sqrt{\frac{kg\,m^3\,kg^{-1}s^{-2}}{m}} = \sqrt{m^2s^{-2}} = m/s \text{ as expected}$$

4.2 Alternative method of calculating escape velocity of Earth

We can approach the same problem from a different perspective. Consider a rocket, mass **m**, on the Earth's surface, then we have

$$mg = \frac{GMm}{R^2}$$

where the Earth has mass **M**, and
where **R** is the radius of the Earth, and
where **G** is the gravitational constant

This gives
$$GMm = mgR^2 \qquad\qquad (A)$$

Now when the rocket is at height x above the Earth's surface, and once again, we are ignoring varying mass for these purposes, we have

$$mg^* = \frac{GMm}{(x+R)^2}$$

where **g*** is the gravitational pull of the Earth at height **x** above the Earth's surface.

But by (A), we have

$$mg^* = \frac{mgR^2}{(x+R)^2}$$

And so, we can conclude that

"According to Newton's Law of Gravitation, the gravitational force on an object of mass m that is a distance x from the surface of the Earth is

$$F = - \frac{mgR^2}{(x+R)^2} "$$

$(\$)$

Now we can give a justification for the Hint in Problem 3.7, namely that $g^* = g \left(\frac{R}{R+x}\right)^2$ where g^* is the gravitational acceleration at height x above sea level, R is the Earth's radius and g is the standard gravitational acceleration at the surface of the Earth.

We have

$$mg = \frac{GMm}{R^2}$$

and

$$mg^* = \frac{GMm}{(x+R)^2}$$

and so, one can easily see that

$$mg\left(\frac{R}{R+x}\right)^2 = \frac{GMm}{R^2} \left(\frac{R}{R+x}\right)^2 = \frac{GMm}{(x+R)^2} = mg^*$$

and so

$$g^* = g \left(\frac{R}{R+x}\right)^2$$

The fact that we are ignoring the changing mass of the rocket for these purposes makes the above relationship between g^* and g an approximation, as noted in Problem 3.7.

Returning to ($\$$), and using Newton's 2nd Law of Motion, we can equate

$$m\frac{dv}{dt} = -\frac{mgR^2}{(x+R)^2}$$

and using the identity

$$\frac{dv}{dt} = \frac{dx}{dt}\frac{dv}{dx} = v\frac{dv}{dx}$$

we obtain

$$mv\frac{dv}{dx} = -\frac{mgR^2}{(x+R)^2}$$

and so, we can solve by separating the variables and integrating both sides

$$\int_{v_0}^{0} v\,dv = \int_{0}^{h} \frac{-gR^2}{(x+R)^2}dx$$

where v_0 is the initial velocity of the rocket and h is the maximum height it attains.

This gives

$$\frac{-v_0{}^2}{2} = \frac{gR^2}{R+h} - \frac{gR^2}{R} = \frac{gR^2}{R+h} - gR = \frac{gR^2}{R+h} - \frac{gR(R+h)}{R+h} = \frac{-gRh}{R+h}$$

Hence

$$v_0 = \sqrt{\frac{2gRh}{R+h}} = \sqrt{\frac{2gR}{\frac{R}{h}+1}}$$

The escape velocity occurs as $h \rightarrow \infty$

i.e. $\frac{R}{h} \rightarrow 0$

and so

$$v_0 \rightarrow \sqrt{2gR} \qquad\qquad (B)$$

But we know that , by (A) above

$$GMm = mgR^2$$

and hence

$$g = \frac{GM}{R^2}$$

Substituting this into (B)

we have

$$v_0 \rightarrow \sqrt{\frac{2GM}{R}} \qquad\qquad \text{just as we derived in 4.1}$$

4.3 Deriving the law of Conservation of Energy from Newton's Second Law of Motion

Regarding the conservation of energy equation as noted above:

Kinetic Energy + Potential Energy = Constant

we can derive this directly from Newton's Second Law as below.

Ignoring any changing mass, we have

$$F = ma = m\frac{d^2x}{dt^2} = m\frac{dv}{dt} = m\frac{dx}{dt}\frac{dv}{dx} = mv\frac{dv}{dx}$$

and so,

$$F = mv\frac{dv}{dx}$$

Now integrate both sides with respect to distance

$$\int F\,dx = \int mv\frac{dv}{dx}\,dx = \int mv\,dv$$

and so

$$\int F\,dx = \frac{1}{2}mv^2 + C$$

Namely,

$$\frac{1}{2}mv^2 - \int F\,dx = constant$$

or

Kinetic Energy + Potential Energy = Constant

This does not represent a 'proof' as such and is more of an intuitive understanding of the concept of the conservation of energy.

It is of interest because it shows the connectedness of mathematical concepts and therein lies more beauty.

Ada Lovelace, 1815 – 1852, said that mathematics should be viewed

"..not merely as a vast body of abstract and immutable truths.."

but that also

"…in their connection together as a whole…"

it was clear that mathematics was the

"….language through which alone we can adequately express the great facts of the natural world, and those unceasing changes of mutual relationship which, visibly or invisibly, consciously or unconsciously to our immediate physical perceptions, are interminably going on…"

And once again, we are back to the beauty, the unimaginably unexpected truth that with just *'pen, paper and brain'*, such things can be precisely mathematically expressed and thereby understood and that, as a result, what once lay hid can now be seen in full light.

5 <u>Sir Isaac Newton and Miss Katherine Storer</u>

"He had lived nearly six years in the same house with Miss Storey [sic], and there is every reason to believe that, during this time, their youthful friendship gradually rose to a stronger passion...."
The Life of Sir Isaac Newton: Containing an Account of his Numerous Inventions and Discoveries, and a Brief Sketch of the History of Astronomy previous to His Time, Compiled from Authentic Documents. [1849] George Grant

The reader will have perhaps surmised that Sir Isaac Newton is this author's mathematical hero.

However, Newton was so much more than a mathematician. His work, of course, encompasses his Laws of Motion and Planetary Motion, including his Law of Universal Gravitation, but also extends to Optics (Newton built the first practical reflecting telescope) and his Theory of Light, with his work on the spectrum , and thereafter to Newtonian fluids and his Law of Cooling. He shares credit with Gottfried Leibniz (1646 – 1716) for the development of the Calculus. But there is much more to his mathematical and scientific legacy – students at school still, for example, learn of the 'Newton-Raphson' method for estimating roots of equations.

He made things, especially when he was young – a mechanical carriage, a water clock and he constructed paper kites, working out the best dimensions and shapes. He made paper lanterns and tied these to his paper kites at night. So great was the effect that his fellow villagers believed them to be comets. This 'practical' side to Newton remained present throughout his entire life and perhaps found its greatest outlet when, as Master of the Mint, for the last thirty of his years, he developed successful anti-counterfeiting methods for coinage.

In his early years, Newton drew and wrote verse, and the walls of his bedroom were covered with his drawings, often charcoal, and sometimes coloured, neatly framed, by Newton himself. Sometimes he copied, but he also drew from life. He drew birds, animals, men, ships, and also mathematical figures, in intricate detail, perhaps showing how important Geometry was to become in his later work.

Newton also studied alchemy, the occult and Biblical chronology. Many have conjectured that he attempted to work out the End of Days by interpreting passages from the Bible, especially from the *Book of Revelation*. Indeed, some commentators have, wrongly, stated that Newton calculated this date definitely to be 2060, largely because of this passage of his writing

"So then the time times & half a time are 42 months or 1260 days or three years & an half, recconing twelve months to a year & 30 days to a month as was done in the Calender of the primitive year. And the days of short lived Beasts being put for the years of [long-]lived kingdoms the period of 1260 days, if dated from the complete conquest of the three kings A.C. 800, will end 2060. It may end later, but I see no reason for its ending sooner."

However, Newton himself wrote

"This I mention not to assert when the time of the end shall be, but to put a stop to the rash conjectures of fanciful men who are frequently predicting the time of the end, and by doing so bring the sacred prophesies into discredit as often as their predictions fail. Christ comes as a thief in the night, and it is not for us to know the times and seasons which God hath put into his own breast."

These studies may seem incongruous to us in these modern times where the distinctions between 'science' and the 'supernatural' are more clear, but in Newton's days, such delineations were more blurry.

There are many excellent and well-researched biographies of Newton and the interested reader is urged to seek them out. This book is not a biography of Newton, although sometimes the author has found himself hearing and sensing something of Newton through the methods of solutions and equations.

It is, however, apposite to touch upon something of the man. And where better to touch upon than his youth.

From the age of around 12 until 17 years of age, he studied at The King's School, Grantham, and, as was common at the time, the syllabus consisted mainly of Latin and Greek, with a little Mathematics.

Whilst there, Newton lodged with a Mr. William Clarke, an apothecary, who lived in a house next to the old George Inn, in the High Street. The George Inn was demolished in 1780, but was rebuilt and remains there still, renamed the George Hotel. Mr. Clarke, it seems, taught Newton in chemistry, and this sparked new avenues of interest for Isaac.

Clarke was the second husband of Katherine Storer, who brought to the marriage four children by her first husband – Arthur, Edward, Katherine and Ann. Together, Katherine and William had two more children named John and Martha

By all accounts Newton was not a distinguished student at first. But then came a motivating factor – Arthur Storer was a bully and had decided to make Newton his target. Eventually Newton had had enough and after one famous incident of bullying, where Arthur Storer had punched Newton in the stomach, the young Isaac, instead of crawling away to nurse his wounds, had picked up Arthur by the scruff of the neck and had smashed his face and nose into a brick wall. And thereafter, Newton decided that another way of beating the bully was to score consistently higher grades than him in all class studies. Arthur was considered one of the brighter students in school and so to Newton, this seemed a good way to beat the bully. Here started 'Newton the Scholar'.

Indeed, Arthur Storer did go on to be academically accomplished, and is regarded as America's first colonial astronomer. Storer's Comet is named after him. Newton references Arthur's work in much of his own but in his well-known confessions list in the Fitzwilliam notebook of 1662 he includes the "beating of Arthur Storer" as one such confession. This list of indiscretions by Newton includes also

"Stealing cherry cobs from Eduard Storer"…. 'Denying that I did so"

But, what of Miss Katherine Storer?

Newton, in all his genius, was also once a boy and a young man. Katherine was one or two years his junior.

Brewster recounts, in his *Life of Sir Isaac Newton* [1829] that Newton preferred the company of Katherine (and her young female companions) to *'all others'* and that he made all manner of things for her – tables and cupboards for her dolls' house, and utensils also, along with numerous trinkets. Brewster also goes on to state that Katherine's *'portion'* (i.e. dowry) was small, and that Newton's *'fortune'* was, at this time, *'inadequate'*, meaning that the consummation of their happiness appears to have been prevented by financial reasons.

Then followed Newton's desire to be a Fellow at Cambridge University. At that time, one could not marry without forfeiting such a chance. Fellows of Cambridge colleges were only allowed to marry from 1860.

It does seem that Isaac and Katherine were star-crossed.

In 1727, Dr Stukeley, antiquarian, physician, Anglican clergyman, friend of Newton, and author of one of the earliest biographies of Newton, visited Katherine Storer, since twice married, aged 82 years, and then known as Mrs. Vincent, and discussed her childhood, and other, memories of Newton.

During those discussions, Katherine told how

'his esteem [for me] never abated during his life'

and

'he never visited Lincolnshire without calling upon [me]'

and that whenever she had sunk into *'pecuniary difficulty'* which seemed to beset her life

'he liberally supplied her wants'

The true meaning, and depth, of this relationship, both for Newton and Miss Storer, are lost in time, but it is evident, that for Newton at least, the feelings lasted a lifetime.

Much has been written about Newton's awkward, often non-existent, relationships with women, but there is evidence enough to point towards a first love between Isaac and Katherine.

After all, Katherine said that

"...[we] were more than friends. He held my hand"

If circumstances had been different maybe Isaac and Katherine would have married and maybe this would have changed the course of Mathematics and Science, but we can never know for sure, because as Newton himself said, the Planets are easier to understand than the vagaries of people:

"I can calculate the motion of heavenly bodies but not the madness of people."

Any errors in this book are mine – please email suggested corrections to **cr648@hotmail.co.uk**

www.ingramcontent.com/pod-product-compliance
Lightning Source LLC
Chambersburg PA
CBHW081518220526
45467CB00010B/2965